PHOTOCHEMISTRY
AN INTRODUCTION

ACADEMIC PRESS RAPID MANUSCRIPT REPRODUCTION

PHOTOCHEMISTRY
AN INTRODUCTION

by
D. R. Arnold
N. C. Baird
J. R. Bolton
J. C. D. Brand
P. W. M. Jacobs
P. de Mayo
W. R. Ware

Photochemistry Unit
University of Western Ontario
London, Ontario, Canada

Academic Press, Inc.   New York and London   *1974*
A Subsidiary of Harcourt Brace Jovanovich, Publishers

ACADEMIC PRESS, INC.
111 Fifth Avenue, New York, New York 10003

United Kingdom Edition published by
ACADEMIC PRESS, INC. (LONDON) LTD.
24/28 Oval Road, London NW1

Library of Congress Cataloging in Publication Data

Main entry under title:

Photochemistry: an introduction.

    1.   Photochemistry.   I.   Arnold, Donald Robert,
Date   [DNLM: 1.   Photochemistry.   QD601.2 P575 1974]
QD708.2.P46   541'.35   74-4442
ISBN 0−12−063350−7

CONTENTS

PREFACE

The past decade has seen remarkable developments in the field of photochemistry. Not only are the fundamental aspects of the science better understood, but there also has been a rapid expansion of our knowledge of the types of photochemical transformations that are possible. Although the synthetic merits of photochemical processes are still to be fully appreciated or exploited, it is only a matter of time until photochemistry takes its place among the essential tools of the synthetic chemist.

This book had its origin in a short course in photochemistry given by the members of the Photochemistry Unit at the University of Western Ontario. The course included topics such as industrial photochemistry and solid state photochemistry, as well as the more conventional introductory material. This departure from the more traditional content of introductory photochemistry courses has been retained in this volume, and the authors hope to provide—by the inclusion of such topics as the spectroscopy and photochemistry of the solid state, industrial applications of photochemistry, photochromism, and a survey of experimental techniques—a book with significant features that distinguish it from other short introductory texts. Because of space limitations, some of the more usual topics have been given only brief mention or have been omitted completely. References have been provided to more detailed and comprehensive discussions of many topics.

CHAPTER 1
INTRODUCTION

A Brief History

Photochemical processes have been intimately related
to the development of Man and his environment even before
his appearance on the planet. It is believed that certain
stages in the generation of the building units for the macromolecules of life occurred on the primordial earth under
the influence of the sun's rays. Subsequently the evolution of the process of photosynthesis, the conversion of
carbon dioxide into carbohydrate, rendered life in its present form possible. Finally the evolution of all life of
an advanced form would be drastically different if the
photochemical process of vision had not been developed.
And the planet continues to be irradiated to the extent of
100 $kcal/cm^2/day$ ...

Photochemical reactions in the laboratory have been
known for almost as long as chemistry has been studied.
Most of the observations were accidental and remained un-
interpreted, and only at the end of the nineteenth century
was any systematic approach made. Then, in Italy, largely
as a result of the work of Ciamician and his collaborator
Silber, and to a lesser extent that of Paterno, the organic
chemist at last paid serious attention to the possibilities
of the chemical action of light. The wide range of reactions discovered by Ciamician and Silber is most impressive; indeed, many of these reactions are still being

studied. In view of this spectacular achievement[1], why
then did the interest in photochemistry decline abruptly?

The immediate reason was the First World War, and its
consequences which rendered continuation of the work difficult for Ciamician and Silber. However, a more permanent
reason was the fact that with the techniques then available
further work was technically difficult. They lacked adequate light sources, filter systems and, more particularly,
physical means of separation of the often complex mixtures.
In addition, chemical and physical theory was in no way
capable of even a partial rationalization of the results
obtained. A good perspective of the state of photochemistry
in 1911 has been provided by Ciamician himself in an address
to the Congress of Applied Chemistry in New York[2]. At a
distance of nearly sixty years his prophetic insight is
truly impressive.

After the First World War photochemistry became the
province of the physical chemist. The photolysis of small
molecules in the gas phase occupied much of the photochemical endeavour for the next 35 years. During this time
essential techniques were evolved both for light sources,
and the obtention of monochromatic beams, and for spectroscopic analysis in general. Methods of chemical analysis
also developed, and, most importantly, theories of chemical
bonding and of quantum mechanics provided the language in
which to speak of the new observations. An intensive **effort**
was devoted to a few supposedly simple processes in an
attempt to understand them. Several hundred papers have
been written on the photolysis of acetone....

---

1. An excellent description of G. Ciamician's achievements
   is given in two papers: M. Pfau and N.D. Heindel, Ann.
   Chim., 187 (1965); idem, J. Chem. Ed., 42, 383 (1965).

2. G. Ciamician, Science, 36, 385 (1912).

In the fifties general interest in photochemistry by the organic chemist again arose. Part of this interest came from an, at first glance, surprising source: natural product chemistry. During the previous fifty years or so a number of natural products, or their derivatives, had been converted into substances of unknown structure by the action of light. The fifties was the heroic age of structural work, and at this time the structures of these photochemical products were elucidated. These were so surprising, being for the most part complex rearrangements, that interest was immediately attracted, and was followed by the deliberate irradiation of natural products which provided readily available complex chromophores.

The sixties saw the emergence of mechanistic organic photochemistry and the merging together of the organic and physical viewpoints. At the present time mechanistic reaction theory may, perhaps, be compared with organic reaction mechanism theory in the 1920's, but new theories, (for instance orbital symmetry relationships) and new techniques (for instance the laser) make for very rapid development.

The Photochemical Reaction[3]

The essential of a photochemical process is that activation for reaction is provided by the absorption of a photon. Photochemical activation differs from thermal activation in that it may be more specific. Light may be absorbed by a particular chromophore which may be a small part of a large molecule, and this process may occur when the molecule is dissolved in vast amounts, relatively, of a solvent.

---

3. A number of texts on photochemistry, particularly organic photochemistry are available. Some of these are listed in the bibliography at the end of this Chapter. In addition there are volumes of reviews. Publications on photochemistry appear in many journals; two in the English language are devoted to the subject: "Photochemistry and Photobiology" and "Molecular Photochemistry".

This leads to the statement that a photon of particular energy, corresponding to a particular wavelength, will only excite a molecule capable of absorbing at that wavelength. This is strictly true for normal light sources, but may require modification for biphotonic processes with intense sources (lasers).

It follows that the only energy available for excitation is that of the photon. If E_2 is the final energy of the system and E_1 that of the molecule in the ground state, then

$$E_2 - E_1 = h\nu$$

where h is Planck's onstant, and ν is the frequency (Hz, formerly $\sec^{-1}$) of the light absorbed.

Expressed in wavelength[4] (λ) or frequency ($\bar{\nu} = \frac{1}{\lambda}$)

we have

$$E_2 - E_1 = \frac{hc}{\lambda} = h\bar{\nu}c$$

For light of 300 nm, for instance

$$\nu = \frac{c}{\lambda} = \frac{3 \times 10^{10} \text{ cm/sec}}{3 \times 10^{-5} \text{ cm}} = 10^{15} \sec^{-1}$$

The expression can be reduced, by inserting values of h and c, and converting units, to:

$$E_2 - E_1 = \frac{2.86 \times 10^{-4}}{\lambda} \text{ kcal mole}^{-1}$$

---

4. For visible and ultraviolet light λ is usually expressed in nm (1 nm = 10^{-9} meter). Formerly this unit was called the millimicron ($m\mu$). One also finds λ expressed in Angstroms (1 Å = 10^{-8} cm).

where λ is in nanometers. Thus 1 mole of photons Einstein) at 300 nm is equivalent to 95.3 kcal mo inspection of Table I it will be seen that light interest to the photochemist comes between the inf _.cu and about 200 nm; and corresponds in energy of from about 40 to 140 kcal mole^{-1}.

TABLE I

Energy Conversion Table

λ(nm)	$\bar{\nu}$(cm^{-1})	kcal mole^{-1}
200	50,000	143.0
250	40,000	114.4
300	33,333	95.3
350	28,571	81.7
400	25,000	71.5
500	20,000	57.2
600	16,666	47.7
700	14,286	40.9

The Quantum Yield

A normal chemical reaction gives a yield of product. In a similar way a molecule absorbing a quantum of light may give a particular product or undergo a particular process with greater or lesser efficiency. This efficiency, the quantum yield (Φ) is defined as:

$$\Phi = \frac{\text{the number of molecules undergoing a process}}{\text{the number of quanta absorbed}}$$

The number of molecules undergoing the process must be measured by some analytical technique (if in any particular case it can be measured at all). The number of quanta

5

absorbed is measured by actinometry (see Chapter 5) which may be chemical or involve a physical device such as a photomultiplier or thermopile.

It is important to be clear as to which process the quantum yield applies, since the actual situation may be complicated. The relationship to the chemical yield may also lead to confusion. Let us take a model, the molecule AB. On absorption of light in solution it is converted into the excited molecule (AB)*. The quantum yield for the generation of (AB)* is necessarily unity if no other parallel process exists. Let us suppose that (AB)* now undergoes homolysis to give the radicals A· and B·. In solution, in a solvent cage, a large proportion of the radicals will recombine to make vibrationally excited AB and the energy will be dissipated as heat. In a matrix, as an extreme, there may be one hundred percent recombination. We then have the following sequence:

$$\Phi = 1 \qquad \Phi = 1$$

$$AB \longrightarrow AB^* \longrightarrow A^· + B^·$$

Each of the photochemical steps has a quantum yield of unity, but the chemical yield of product is zero.

A reverse situation may obtain. In the scheme below the molecule A is excited to A*. The quantum

$$\Phi = 1 \qquad \Phi = 0.001$$

$$A \longrightarrow A^* \longrightarrow B$$

yield for formation of B is very low (0.001); only 1 in every thousand excited molecules giving product. The rest of the excited molecules decay back to A. If, however, there is no other competing process, given enough time and light, all A will be converted to B. The chemical yield may thus approach 100% when the quantum yield is 0.001.

It may be noted parenthetically that if one molecule of product is the maximum that can be obtained from each excited molecule a quantum yield of unity is the theoretical maximum. Quantum yields greater than unity have, however, been found. They necessarily imply that a chain process involving non-photochemical steps is involved in the events subsequent to the primary photochemical act.

What are the possibilities for behaviour which confront the photochemically excited molecule? These are represented in the following diagram:

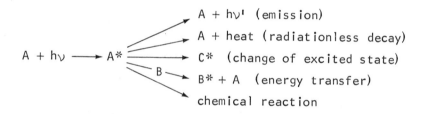

Before briefly considering the nature of these various possibilities, it is necessary to discuss the state of the excited molecule, A*.

Electronic Transitions

An atom or molecule can exist only in discrete energy states. These states are characterized, in quantum mechanics, by wave functions, ψ, which are solutions of the Schrodinger equation. The wave function defines the orbitals and properties of electrons in molecules. An inexact but useful pictorial idea is that of the one-electron orbital whose probability density at any point in space is given by ψ^2. Orbitals are commonly represented pictorially by a surface of points of equal electron density such that a major fraction (>90%) of the charge is inside the volume defined by the surface. These orbitals may be localized on one atom or delocalized over two or more nuclei (a molecular orbital). Each orbital may contain no more than two electrons and then these must have opposite spins.

In the combination of two identical atomic orbitals two molecular orbitals result. One of these is of lower energy than the atomic orbitals involved and one is higher. As a result of the lower energy the atoms are bonded. If one electron is donated by each constituent atom this bonding MO contains the permitted two electrons of opposite spin.

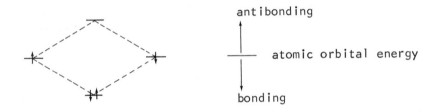

antibonding

atomic orbital energy

bonding

Addition of a further electron requires that it go into the higher energy orbital. Since this is of higher energy than the constituent atomic orbitals it is anti-bonding. For thermal processes of chemical substances or intermediates involving the first row of elements the antibonding orbitals are normally not used. This is not the case in photochemistry.

There are three classes of molecular orbitals with which we shall be concerned[5]:   n, π and σ.

---

5.   H. H. Jaffe and M. Orchin, "Theory and Application of Ultraviolet Spectroscopy", Wiley, New York, 1962, provides a good coverage for this and following sections.

8

n Orbitals

These are the lone pairs of electrons situated on hetero-
atoms. In certain cases they may be called <u>non-bonding</u>
because, to a first approximation they take little part
in the bonding process. Such is the case with the p
orbital on the carbonyl oxygen which is not part of the
double bond, but is at right angles to it. A similar
situation obtains in pyridine.

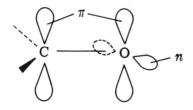
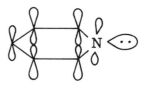

In pyrrol, however, the lone pair electrons are involved
in the aromatic system and so do contribute to the overall
bonding. There are thus two types of n orbitals; both are
characterized by a low ionization potential.

π and π* Orbitals

The π orbitals (and the corresponding π* antibonding
orbitals) are usually formed in systems containing the
first row elements by overlap of p orbitals. These are
illustrated for the carbonyl group. Here the greater
electronegativity of oxygen causes a displacement of
electron density from that found in the symmetrical
ethylene.

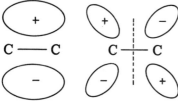
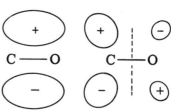

Bonding   Antibonding   Bonding   Antibonding

ETHYLENE   CARBONYL

9

The π, π^* orbital system, simple, as in ethylene or the carbonyl group, or more complex, as in butadiene or benzene, is that which most concerns the organic chemist.

σ and σ^* Orbitals

The σ orbitals constitute the framework of organic molecules and are of very low energy, whereas the σ^* antibonding orbitals are of comparatively high energy. To a first approximation they are not involved in most reactions with which we shall be concerned. The orbital is cylindrically symmetrical about the axis joining the two centers.

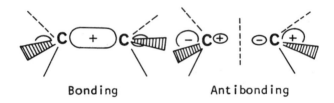

Bonding   Antibonding

In our descriptions of the states of molecules these orbitals will be largely ignored.

The electronic configuration of a molecule is obtained by adding electrons to molecular orbitals. For instance the ground state of formaldehyde is represented:

$$S_o = (1s_0)^2(1s_C)^2(2s_0)^2(\sigma_{CH})^2(\sigma_{CH'})^2(\sigma_{CO})^2(\pi_{CO})^2(n_0)^2$$
$$(\pi^*_{CO})^0(\sigma^*_{CO})^0$$

where the superscript indicates the number of electrons in the orbital. However, ignoring the low or high energy orbitals this is simplified to:

$$S_o = (\pi_{CO})^2(n_0)^2(\pi^*_{CO})^0$$

10

Photochemical excitation involves the transfer of an electron from a lower orbital into a higher one. Thus, in the case of formaldehyde an n electron may be excited into the vacant π^* orbital. Such may be designated an $n \to \pi^*$ (or $\pi^* \leftarrow n$) transition. The final state is called (n, π^*). Similarly a π electron may be excited into a π^* orbital. This is a $\pi \to \pi^*$ transition and the final state is (π, π^*).

Some transitions and configurations of the related excited state are shown below for a carbonyl compound:

State	Transition	Configuration
$n\pi^*$	$n \to \pi^*$	$(\pi_{CO})^2 (n_0)^1 (\pi_{CO}{}^*)^1 (\sigma_{CO}{}^*)^0$
$n\sigma^*$	$n \to \sigma^*$	$(\pi_{CO})^2 (n_0)^1 (\pi_{CO}{}^*)^0 (\sigma_{CO}{}^*)^1$
π, π^*	$\pi \to \pi^*$	$(\pi_{CO})^1 (n_0)^2 (\pi_{CO}{}^*)^1 (\sigma_{CO}{}^*)^0$

In the excited state occupied molecular orbitals are not completely filled. It is thus possible that the electrons originally (antiparallel) in the ground state may be parallel or antiparallel. If antiparallel then the molecule is in a singlet state (represented by S) as it was the ground state; if parallel then the state is a triplet (T). The distinction is indicated by the superscript preceding the state description where a one indicates a singlet and a three a triplet.

This extra information may be indicated in the simple representation of the configuration, but not very elegantly, as shown below:

State	Transition	Configuration
$^1(n, \pi^*)$	$n \to \pi^*$	$(\pi_{CO})^2 (n_0\uparrow)^1 (\pi_{CO}{}^*\downarrow)^1 = S_1$
$^3(n, \pi^*)$	$n \to \pi^*$	$(\pi_{CO})^2 (n_0\uparrow)^1 (\pi_{CO}{}^*\uparrow)^1 = T_1$
$^1(\pi, \pi^*)$	$\pi \to \pi^*$	$(\pi_{CO}\uparrow)^1 (n_0)^2 (\pi_{CO}{}^*\downarrow)^1 = S_2$

Fluorescence and Phosphorescence[6]

The excited state of a molecule is metastable. One method by which it may decay is by the emission of light, usually of somewhat longer wavelength than that absorbed. If the emitting species is a singlet then the emission is called fluorescence; if a triplet, then it is termed phosphorescence. Since the ground state is obtained in both cases it will seem that in the latter case there is a change of spin in the overall process. More strictly, fluorescence is emission without change of multiplicity and phosphorescence is emission with such a change.

In the ground state most molecules, following a Boltzmann distribution of energies, are in the lowest vibrational level. On excitation, subject to certain constraints, various vibrational levels are initially populated. This is shown below diagramatically (A) for one vibrational mode only. The probability of transition between the vibrational levels is not equal for each transition. If one represents this relative probability by the height of a vertical line then the absorption spectrum (B) is obtained. Of course most molecules encountered by photochemists have very many vibrational modes and the resulting envelope which results contains a large number of transitions.

Very rapidly ($\sim 10^{-12}$ sec) the vibrationally excited molecules in the electronic excited state are deactivated, particularly in solution, to a Boltzmann distribution of energies in which the lowest vibrational levels of the excited state are most populated. Then emission occurs at a rate depending on the environment and the particular type of molecule. The state reached after emission is the ground state, but, with restrictions, at various excited vibrational energy levels (C). Again representing the probability

---

6. R. Becker, "Theory and Interpretation of Fluorescence and Phosphorescence", Wiley Interscience, New York, 1969.

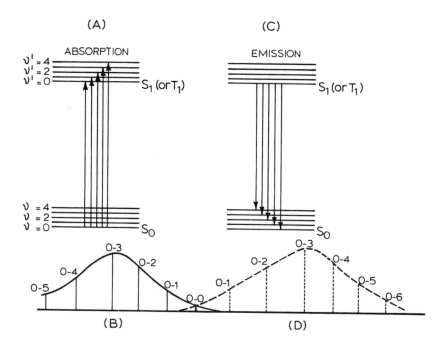

Figure 1   Energy level diagrams and spectra illustrating the displacement of an emission maximum to longer wavelengths relative to the absorption maximum.

of a transition by a vertical line we obtain (D) the emission spectrum. The technique by which the measurements are made is described in Chapter 5.

Generally, one transition is in common - that between zero vibrational levels. This comes at lowest energy in the absorption spectrum and at highest energy in the emission spectrum. This 0-0 band, as it is termed, indicates the maximum energy available from an excited molecule. It should be noted that such a number cannot always be obtained spectroscopically since not all compounds emit either fluorescence or phosphorescence and many emit neither;

and many spectra are structureless so that the transition cannot be recognized. In addition, the 0-0 transition may be forbidden.

It might also be thought that emission from higher electronic states might also be observed. With rare exception this is not so. Radiationless decay by other means occurs rapidly and whatever the excitation, it is the lowest excited state, in the singlet and/or triplet manifold, which emits.

It will be noted that the 0-0 band is not always the most intense. It may, in fact, be so weak as to be undetectable. This is a consequence of the fact that nuclear movements are very slow by comparison with electronic motion. Essentially the nuclei remain stationary in the very short period - of the order of $\frac{1}{\nu} \approx 10^{-15}$ sec - required for the absorption of the light quantum. The qualitative statement of the <u>Franck-Condon principle</u> implies that <u>vertical</u> transitions, that is, transitions which take place without change in geometrical structure or momentum of the molecule are the most probable.

How this may show itself in the shape of a spectrum may be seen from inspection of potential-energy diagrams for a hypothetical diatomic molecule. In curve A, Figure 2, excitation from the lowest vibrational level vertically - that is without change in r, the internuclear distance - leads to the lowest vibrational levels of the excited state.

Here the Franck-Condon principle predicts that this transition is probable. The fact that vertical excitation (no change in r) does lead to the vibrationally unexcited electronically excited state implies that the excited state does not differ very much in relaxed geometry from the ground state. The same consequences apply to the reverse process, emission. In this case the 0-0 band will be relatively strong.

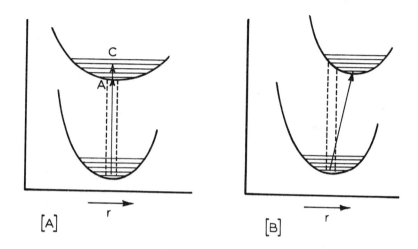

Figure 2   Potential energy diagrams illustrating no
change (A) or change (B) in the equilibrium
configuration between an upper and lower state.

In curve B vertical excitation leads to a higher vibrational level of the excited state. Excitation to the
zero vibrational level requires a change in r. The Franck-
Condon Principle states that this is unlikely. This implies
that the geometry of the relaxed excited state is different
from that of the ground state, and so the 0-0 band is weak.

Excitation takes place to the envelope of the potential energy curve. Here, as in the ground state, kinetic
energy is zero (a turning point in the respective vibration). Excitation to a point not on the envelope such as
point C (Curve A) means that some potential energy has been
converted to kinetic energy, a process which during excitation is Franck-Condon forbidden since it implies a change
in momentum.

Radiationless Transitions

The above discussion has been concerned with electronic transition involving radiation. It is also possible for electronic transitions to occur without concomitant radiation.

Since the excited molecules with which we will be concerned are of two types, singlets and triplets, we are concerned with two types of radiationless transitions. The first, internal conversion does not involve any change in multiplicity in the transition. Thus $S_1 \rightsquigarrow S_0$, $S_3 \rightsquigarrow S_2$ where S_2, S_3 etc. are higher energy excited singlets, are examples of such a transition. Note that a wiggly arrow is used to indicate a process not involving radiation

The second process is termed intersystem crossing, and here a change in multiplicity is occurring. Examples include $S_1 \rightsquigarrow T_1$ and $T_1 \rightsquigarrow S_0$. There are quantum mechanical rules which govern the probability of such transitions. Of these the most important restriction is that of spin which may decrease the rate of such a process by a factor as high as 10^6 over the equivalent singlet process.

The rates of internal conversion (i.e. change of state without change of multiplicity) amongst excited singlet states are of the order of $10^{12} sec^{-1}$ and are generally faster than the rate of fluorescence. In the same way higher triplets most probably decay to the lowest triplet at similar rates. The process may be viewed as an iso-energetic one where a higher vibrational level of a lower excited state couples with a low vibrational level of a higher excited state. This is represented in Figure 3 (known as a Jablonski diagram) by a horizontal wiggly line. Thermal equilibrium with the environment then occurs (vibrational cascade), resulting in the usual Boltzman distribution of vibrational energy.

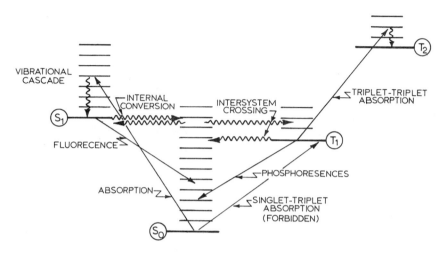

VIBRATIONAL
CASCADE

INTERNAL
CONVERSION

INTERSYSTEM
CROSSING

TRIPLET-TRIPLET
ABSORPTION

S_1

FLUORECENCE

T_1

PHOSPHORESENCES

ABSORPTION

SINGLET-TRIPLET
ABSORPTION
(FORBIDDEN)

S_0

T_2

Figure 3   Energy level diagram indicating the principal
light induced photophysical processes.

The decay through the higher excited states is rapid,
as mentioned, probably because of the higher density of
excited states and the increased probability of coupling.
The final jump down, $S_1 \rightarrow S_0$, which has the largest energy
gap, and in which the density of states is lowest, is far
slower. As a result fluorescence (radiative decay) or
intersystem crossing (i.e. change of state with change of
multiplicity) to T_1 can compete effectively with the radiationless processes.

The radiationless transition $S_1 \rightsquigarrow T_1$ may also be
represented using potential energy curves, as shown in
Figure 4. In the diagram the molecule is excited to S_1.
Vertical (the most probable) excitation generates S_1 in a
higher vibrational level (X). This is followed by dissi-
pation of the excess vibrational energy until the level Y is
reached. At this energy the potential energy curves for
singlet and triplet intersect. At point Z all the energy is
contained as potential energy and "crossing" from S_1 to T_1
can occur.

17

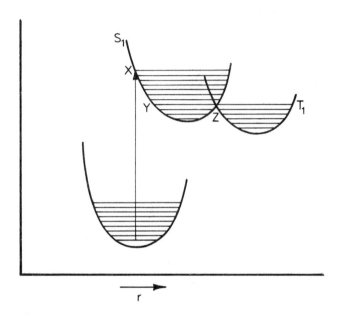

Figure 4   A rough schematic representation of the process $S_O + h\nu \rightarrow S_1 \rightarrow T_1$.

Lifetimes of Excited States

From the foregoing it may be seen that the lifetime of an excited state is a consequence of the rates of the various possibilities open to it.   There are a number of terms which require definition.

The excitation from S_O to S_1 may be subject to few restrictions.   This will be reflected in a high extinction coefficient (ca $10^4 - 10^5$).   The reverse radiative process, fluorescence, must be equally allowed.   The radiative lifetime, τ_o, defined as the reciprocal of the unimolecular rate constant for fluorescence (k_f) can be related, in fact, directly to the integrated intensity of the absorption band in the reverse process.

However, the lifetime, experimentally, of the excited state will only correspond to τ_0 if no other process, unimolecular or other, interferes. The experimentally determined lifetime (Chapter 5) τ is the reciprocal of the sum of <u>all</u> processes open to the excited molecule. These will include all bimolecular processes as well as the unimolecular ones of internal conversion and intersystem crossing, and chemical transformation. If restrictions, such as those of spin, decrease the "allowedness" of these processes, the excited molecule may have a surprisingly long lifetime. In solution in practice it is virtually impossible to remove interfering impurities - oxygen is probably the most serious - which introduce bimolecular (or pseudo unimolecular) decay processes. In a solvent matrix at low temperature (that of liquid nitrogen) or in a plastic matrix this effect, and others, such as self-quenching, are minimized. Under these conditions some aromatic hydrocarbons have triplet lifetimes of the order of seconds. This is to be contrasted with singlet lifetimes which may be as short at 10^{-10} sec. and rarely longer than 10^{-6} sec.

It should be remembered that intramolecular chemical processes, both for the singlet and the triplet, which lead to isomerized molecules, may represent mechanisms by which the molecule disposes of its excess energy. If the product of such a transformation is very unstable and reverts, thermally, to the original ground state molecule, its presence may not be detected. Such appears to be the case in the photochemistry of benzene, for instance.

Spin-Orbital Coupling

It has been mentioned that excitation and emission processes are "improbable" when a change of spin is involved. The consequences are a very low extinction coefficient for absorption $(S_0 \rightarrow T_1)$, which may render the transition undetectable, and produce a very long lifetime for emission (phosphorescence). This inefficiency occurs because transitions between pure singlet and triplet states are quantum mechanically forbidden. For the transition to occur at all some perturbation, internal, external or both is required. In internal perturbation, the (low) intensity of singlet → triplet transition is determined by how much singlet character is mixed by spin orbit coupling into the triplet state. The coupling is more pronounced when the electron is in a stronger field, that is when the electron is in an orbital which has a higher probability of being close to the nucleus, especially if that nucleus is of higher atomic number. The coupling is also more pronounced if the energies of the states being mixed are close.

Intramolecular spin-orbit perturbation may be enhanced by the attachment of a heavy atom to the molecule concerned It might then be expected that the processes $S_1 \rightsquigarrow T_1$, $T_1 \longrightarrow S_O$ and $T_1 \rightsquigarrow S_O$ would all be enhanced since they are all intercombinational i.e. involve spin change. One could expect a greater yield of triplets, a greater yield of phosphorescence (as compared with fluorescence) and a shorter phosphorescence lifetime. This is illustrated (in terms of phosphorescence lifetimes) for substituted naphthalenes in the following table:

TABLE 11
Phosphorescence Lifetimes in Seconds

	Solvents			
	EPA	PrCl	PrBr	PrI
Naphthalene	2.5	0.52	0.14	0.076
1-Fluoronaphthalene	1.4	0.17	0.10	0.029
1-Chloronaphthalene	0.23	0.075	0.06	0.023
1-Bromonaphthalene	0.014	0.007	0.007	0.006
1-Iodonaphthalene	0.0023	0.001	0.001	0.001

[EPA is a mixture of alcohol, isopentane and ether]

It can be seen also from the same Table that perturbation can occur in an intermolecular way if the heavy atom is contained in the solvent. It can be shown that the effects are not chemical, because a similar series can be obtained using the rare gases as solvents.

The energies of the relevant states are also important. The $S_1 \rightsquigarrow T_1$ transition is dominating in carbonyl compounds and they thus do not, for the most part, fluoresce. This is in sharp contrast with aromatic hydrocarbons, and it is important for the understanding of the chemistry of the carbonyl group. One reason is that the singlet-triplet $(S_1 - T_1)$ splitting is small in ketones - usually near 5 kcal/mole, but is much larger in benzene ($\sim$ 30 kcal/mole). In addition the π bonds in benzene are such that the electrons therein are not close to the nucleus. The n orbital of the carbonyl group, containing some s character, allows the contained electrons to spend more time near the nucleus, i.e. under greater perturbation.

There are other perturbations which may affect intercombination. Amongst these the most important are the influence of impurities, notably oxygen, and the formation of complexes such as change-transfer complexes. The latter are loose molecular associations formed by transfer of charge between the components and electrostatic bonding.

Energy Transfer[7]

This represents another channel whereby the excited molecule can dispose of its energy. Rather than do anything itself in the time available it may simply give its energy away. As one might suspect from what has gone before, there are restrictions on this process also.

The concept of energy transfer can be represented in the following very general equation:

$$A + D^* \rightarrow D + A^*$$

7. A.A. Lamola in "Energy Transfer and Organic Photo-chemistry", Technique of Organic Chemistry, Volume XIV, Interscience, 1969, p. 17 et seq.

where the asterisk indicates an excited molecule. The process can be described in two complementary ways depending on one's point of interest. If the object is to remove the energy from D^* (Donor) then the process can be described as quenching and A (Acceptor) is the quencher.

On the other hand, it might be the object to generate A^* indirectly rather than by the irradiation of A. This might be the case if the particular excited state of A (A^*) was inefficiently formed by direct irradiation. The formation of A^* indirectly may then be said to be sensitized.

For the transfer of energy to occur, by whatever process, we might expect that A^* should be lower in energy than D^*. One might imagine transfer taking place isothermally, but then one would be faced with the problem that back transfer of energy should be equally possible. In the extreme, slow transfer might even take place endothermally if the deficit could be made up by the Boltzman distribution in vibrational levels. But now the back transfer should be much faster. It is thus important, in planning such experiments, to know the energies of the states involved. Conversely, such experiments may be used to investigate such energies.

It is also an important requirement that the transfer of energy take place within the excited lifetime of the donor. Depending on the particular mechanism of energy transfer, this dictates the minimum concentration of the acceptor necessary for effective transfer.

At present there are believed to be two general types of energy transfer processes and one 'trivial' process. Since the latter is disposed of quickly it will be dealt with first.

The trivial mechanism can be represented by the following equation:

$$D + h\nu \rightarrow D^*$$
$$D^* \rightarrow D + h\nu'$$
$$A + h\nu' \rightarrow A^*$$

22

- that is, the <u>light</u> emitted by the donor travels through the solution and is absorbed by the acceptor. This is essentially the direct excitation of A by absorption of radiant energy. The absorption of the energy by A has no effect on the lifetime of D^*; such is not the case for true energy transfer.

For energy transfer to take place the excited molecule which after some $10^{-11} \sim 10^{-12}$ sec has a Boltzmann distribution of vibrational energies in the excited state, must have available to it an isoenergetic transition in a neighboring molecule. Then if coupling does occur, removal of the energy from D^* occurs simultaneously with its appearance in A. This is illustrated in Figure 5.

Transfer may take place where two molecules approach each other closely, provided energy considerations are favorable. The distances involved are those approaching molecular contact, and hence the rate of energy transfer by this, the <u>exchange mechanism,</u> is limited by the diffusion

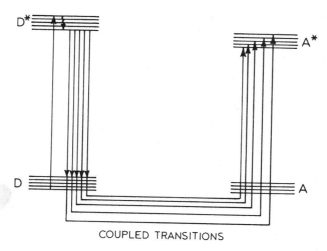

COUPLED TRANSITIONS

Figure 5   Pictorial representation of the process $D^* + A \rightarrow A^* + D$.

rates of the molecules in solution (until such a concentration as all the excited molecules have a nearest neighbor acceptor). This implies that the rate of energy transfer will show a marked viscosity dependence.

When the molecules D^* and A are close the electron
clouds penetrate one another. In the overlap region the
electrons cannot be properly described as belonging to one
molecule rather than the other. This includes the excited
electron also. The quantum mechanical restrictions on this
process are less stringent than on the second method of
energy transfer described below. In particular the spin
function of D^* or of A^* need not be equal to that of the
ground state molecule.

Transfer may also occur between molecules which are
much more widely separated and here viscosity will be much
less important. The distances involved may be up to 50 -
100A. In this process energy matching is also required for
transfer. This mechanism is termed Resonance-Excitation
energy transfer.

The system may be approximated classically by the
coupling of two mechanical oscillators such as two tuning
forks on the same base or two swinging pendulums on a connecting string. Here the oscillating electric charges
interact as two dipoles. The interaction is strongest if
the corresponding dipole transitions in D^* emitting and in
A absorbing are not spin forbidden i.e. are singlet-singlet
or triplet-triplet. The dipole-dipole energy is dependent
on the inverse third power of the distance between molecule
and the probability of energy transfer falls off as the inverse square of the interaction energy. Hence the rate of
transfer decreases to the sixth power of the intermolecular
distance.

From the point of view of the organic photochemist
exchange energy transfer mechanism has a particular virtue.
It permits the following process to occur.

$$D^*_T + A \rightarrow A^*_T + D$$

If the triplet of A (A_T^*) is of difficult access, the above process using another triplet (D_T^*) may be the only practical route to its generation. In recent years this technique has been widely used to 'sensitize' triplet chemical reactions of dienes and alkenes, substances in which intersystem crossing is inefficient. The method is best understood by the consideration of a particular example, which is illustrated in Figure 6.

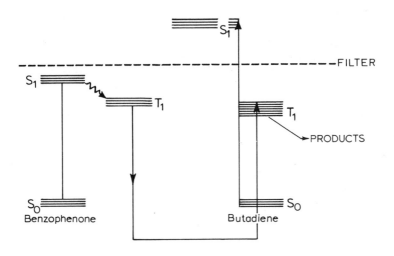

Figure 6   Triplet-triplet energy transfer.

As the triplet donor a substance is required in which intersystem crossing is very efficient and which does not undergo intramolecular chemical change. The aromatic ketones fulfil these requirements. Irradiation of benzophenone gives the singlet which is, with a quantum yield of unity, converted to the lowest triplet (of 69 kcal/mole energy) which is (n, π^*) in character. Let us assume that the acceptor is butadiene. This has a triplet energy in the region of 55 kcal. Energy transfer is thus possible. The experiment is conducted as follows. A solution of benzophenone and butadiene is irradiated. Butadiene does not absorb at wavelengths longer than 290 nm so if a pyrex filter is interposed none of the incident light is absorbed by the butadiene. (The $S_0 \to T_1$ absorbance is far too low to interfere). Light is, however, absorbed by the benzophenone which passes, as described, into the triplet state. Then, if the concentration of butadiene is sufficient to ensure collision before decay of the benzophenone triplet, efficient energy transfer will occur with the generation of the butadiene triplet.

Collisional transfer is also possible between singlets and between triplets separately. The latter may seem at first glance surprising since the concentration of triplets is normally so small that one would not expect a collision between two of these. However two long-lived triplets may collide, under suitable conditions, and annihilate each other. A singlet is the usual result of this process which is termed triplet-triplet annihilation.

$$D_T^* + D_T^* \to D_S^* + D$$

$$D_S^* \to D + h\nu$$

The overall process may be detected because of the observed fluorescence of D.

This does not exhaust the catalogue of possibilities: in particular, mention should be made of the fact that the two components of the energy transfer system may be contained in the same molecule.

We have now covered, albeit very briefly, the possibilities of physical behavior open to the excited molecule. Some of these will be explored in greater detail in the subsequent chapters. Only one mode of energy utilization has been ignored: chemical reaction. This will form the backbone of most of the later chapters, but we will consider here some of the primary photochemical processes which are available to the excited molecule.

Primary Photochemical Processes

The variety of first-order and second-order primary photochemical processes which can occur are illustrated below.   Each of these will be considered in turn.

First-order processes

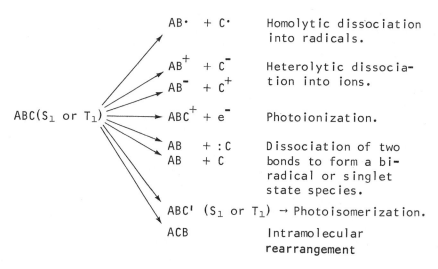

ABC(S_1 or T_1)

AB· + C·   Homolytic dissociation into radicals.

AB$^+$ + C$^-$ Heterolytic dissocia-
AB$^-$ + C$^+$ tion into ions.

ABC$^+$ + e$^-$ Photoionization.

AB + :C   Dissociation of two
AB + C   bonds to form a bi-
             radical or singlet
             state species.

ABC' (S_1 or T_1) → Photoisomerization.

ACB   Intramolecular
             rearrangement

Second-order processes

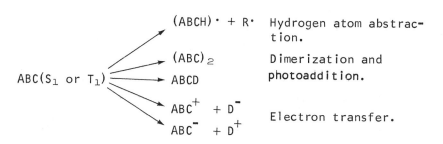

ABC(S_1 or T_1)

(ABCH)· + R·   Hydrogen atom abstrac-
               tion.

(ABC)$_2$ Dimerization and
ABCD   photoaddition.

ABC$^+$ + D$^-$
ABC$^-$ + D$^+$ Electron transfer.

27

Homolytic Dissociation into Radicals

This is one of the most common types of primary photo-chemical processes as it only involves a simple unimole-cular homolytic bond cleavage. Figure 7 illustrates three ways in which this might occur for the diatomic molecule AB. Excitation into a repulsive state (a continuum part of the absorption spectrum) will result in dissociation either to ground state atoms or with both atoms electronically ex-cited (Figure 7a). Excitation to a bound state (discrete part of the absorption spectrum) followed by crossing to a repulsive state will also result in dissociation (Figure 7b). This process is called predissociation and may involve an activation energy if the crossover point lies much above the ground vibrational level of the bound excited state. Dis-sociation may also result from excitation to a bound state if the molecule ends up at a point above the dissociation limit (Figure 7c).

The above considerations also apply to polyatomic mole-cules, although it may not be possible to ascertain the detailed mechanism of the dissociation as a knowledge of the nature of all the excited states may not be available.

As an example of a dissociation into free radicals, consider the Norrish "Type 1" reaction in ketones:

$$CH_3COC_2H_5 \xrightarrow{\quad (1) \quad} CH_3CO^{\cdot} + C_2H_5^{\cdot}$$
$$\xrightarrow{\quad (2) \quad} CH_3^{\cdot} + C_2H_5CO^{\cdot}$$

Towards the long-wavelength side of the absorption band, there is a tendency for the weakest bond to break (e.g. reaction (1) above); however, as the excitation proceeds to shorter wavelengths the alternative process involving cleavage of a stronger bond may also occur (e.g. reaction (2) above).

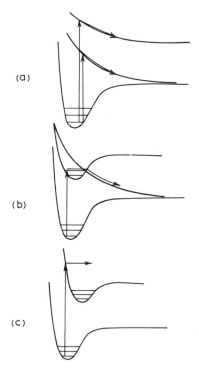

Figure 7   Three paths for a photochemical dissociation
of the diatomic molecule AB. (a) Direct
excitation to a repulsive state (b) Excitation to a bound state followed by intersystem crossing to a repulsive state (predissociation) (c) Excitation to a bound
state but above the dissociation limit.

Heterolytic Dissociation into Ions

This process is not very common and only occurs when there is a large difference in the electronegativity of the two fragments.

Photoionization

No bound state is involved in this process as here the excitation energy exceeds the first ionization potential of the molecule producing a free electron and the positive ion of the original molecule. As this process usually requires 8-10 eV of energy, it is not an important process except in the vacuum ultraviolet. For example at $\sim$110 nm acetylene will photoionize.

Dissociation of Two Bonds

There is a wide variety of reactions which involve dissociation of two bonds as a primary step. This may occur in a concerted fashion or stepwise. In some cases it is possible to tell the difference from the stereospecific nature of the products.

One of the simplest reactions of this type is the photodissociation of diazarene at $\sim$313 nm

$$CH_2 \diagup\!\!\!\overset{N}{\underset{N}{\|}}\!\!\!\diagdown + \ h\nu \ \longrightarrow \ CH_2(S) \ + \ N_2$$

Singlet methylene is almost exclusively formed; however, conversion to triplet methylene occurs on practically every collision.

Photoisomerization, rearrangement and cycloaddition

This group comprises a major fraction of organic photochemical reactions, and will be the subject of two later chapters.

The simplest transformation is probably <u>cis-trans</u> isomerization about a double bond

Rearrangements may involve electrocyclic processes such as

in which sigma bonds are formed (or lost) at the expense of π-bonds. Such reactions are governed by orbital symmetry consideration. Cycloadditions, if concerted are also under such control, but discrete intermediates may also occur, in which case the primary step resembles a radical addition. Sigmatropic reactions are reactions where the numbers of π and σ bonds remain constant but are differently distributed in the product from the starting material.

CHAPTER 2

THEORY AND THE EXCITED STATE

Introduction

Although chemists believe that the quantum-mechanical description of nature is a sufficiently good model for most chemical phenomena, it is clearly recognized that formidable mathematical difficulties are involved in applying quantum mechanics to real chemical systems. For this reason, the story of quantum chemistry in the last forty years has involved the development and application of <u>approximate</u> quantum mechanical methods. Most chemists are familiar with the manner in which such techniques are applied to describe and predict the bonding, structure, energetics, and reactivity of the ground states of molecules; in the present chapter the application of these models to describe the same properties for molecules in their <u>electronically excited states</u> is considered. It should also be mentioned that quantum mechanics can also be employed to study the inter-conversion of states, whether by radiative or non-radiative processes, although this topic will not be considered further in this chapter[1,2].

---

1. See however the remarks in Chapters 1 and 3.

2. For a review of current theories regarding radiationless transitions, see (a) B.R. Henry and M. Kasha in <u>Ann. Rev. Phys. Chem.</u> 19, 161 (1968) and (b) J. Jortner, S.A. Rice and R.M. Hochstrasser in Vol. 7, "Advances in Photochemistry", Ed. by J.N. Pitts, W.A. Noyes Jr. and G. Hammond, Wiley, New York, 1969.

Bonding in Electronically-Excited States

Although both the valence-bond and molecular-orbital approximations are commonly employed in dealing with the ground electronic state of molecules, most theoretical investigations of excited states employ the latter theory. For this reason, the molecular orbital (MO) theory will be used throughout this discussion[3].

Basic to the MO theory is the concept that one can classify each of the individual orbitals of a molecule as either bonding (BMO), nonbonding (NBMO), or antibonding (ABMO). For present purposes, these orbital types can be distinguished by considering the effect to the overall bonding of adding an electron e^- to the positive ion A^+ of a molecule. Although the addition of the electron invariably corresponds to an exothermic process, it does not necessarily follow that the new electron stabilized the bonding. If the bonding in neutral A is stronger than in A^+, then the orbital occupied by the additional electron is defined as bonding. In contrast, if the additional electron destabilizes the bonding, the orbital it occupies is termed antibonding. The latter case is not as uncommon as one might suspect, particularly if A is a diatomic or triatomic molecule.

The distribution of electrons into bonding and antibonding levels in diatomics can be deduced in a very simple manner from a diagram of the type shown in Figure 1 - here each molecular orbital in the diatomic AB is correlated with the atomic orbitals of free A and free B. Each pair of AOs (e.g. 2s of A with 2s of B) yields two molecular orbitals, one of which is bonding, and the other antibonding. The MOs of AB are filled, in order and according to Hund's rule, in

---

3.   Most of the fundamental work on the molecular orbital theory was developed by R.S. Mulliken, for which he has received a Nobel Prize.

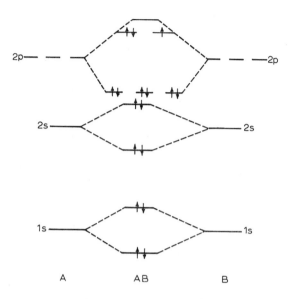

Figure 1   Molecular orbitals in F_2^+

the same manner as used in deducing electron configurations
for atoms. (Note that the positions of the three bonding
MOs formed from 2p AOs are not exactly degenerate as the
diagram implies; the exact ordering of the three does
change with the nuclear charges and number of electrons,
however, so no one diagram can represent their relative
positions correctly for all systems). The explicit electron
configuration illustrated in Figure 1 corresponds to the F_2^+
system; it is obvious that any additional electron must
enter an antibonding MO, and thereby reduce the net bonding.

For polyatomic molecules, nonbonding molecular orbitals
are possible in addition to BMOs and ABMOs. Electrons
placed in NBMOs do not significantly alter the overall bonding. Examples of NBMOs are i) the orbitals occupied by
the valence "lone pairs" on oxygen in formaldehyde, and ii)
the singly-occupied π MO in the allyl radical.

The classification of orbitals in polyatomic molecules is complicated by the fact that while a particular MO is net bonding (or antibonding), it can be antibonding (or bonding) in certain <u>regions</u> of the molecule. For example, the orbital an electron would occupy when added to the butadiene cation $C_4H_6^+$ is <u>net</u> bonding; however this overall effect is due to the fact that the stabilization it imparts to the C_1-C_2 and C_3-C_4 bonds is greater than its destabilization of the C_2-C_3 bond. Thus one would expect that the 1-2 and 3-4 bonds in C_4H_6 are shorter (stronger) than in $C_4H_6^+$, but that the 2-3 bond becomes longer (weaker).

From the examples discussed above, one can surmise that the prediction of the bonding, energy, shape and reactivity for electronically-excited molecules is not a trivial matter; the differences in the behavior between the ground state and the excited state of any molecule will depend upon which bonds are strengthened and which are weakened when an electron is "excited" from one orbital to another. In the realm of photochemistry, the job of theory is to predict the detailed bonding and antibonding characteristics of these orbitals.

As an example of an energy-level diagram for a polyatomic molecule, the energies and bonding character for the molecular orbitals of formaldehyde[4] are illustrated in Figure 2. The two lowest-lying MOs are clearly associated with 1s "inner shell" electrons on oxygen and carbon. Lying between this pair of orbitals and the five MOs clustered in the -10 to -23 e.v. range is an orbital which is mainly a 2s "oxygen lone pair" orbital. The next three most stable MOs are all of the σ type and either C-H or C-C bonding in character. The highest two occupied orbitals are the C-O π-bonding MO and the nonbonding oxygen 2p lone pair, the latter being the least stable to be occupied in the ground state. All four antibonding "valence-shell" orbitals are of positive energy; as expected, the π^* MO of the C-O group is significantly more stable than are the three σ^* MOs.

---

4.   From an unpublished "<u>ab initio</u>" calculation by J.R. Swenson and N.C. Baird.

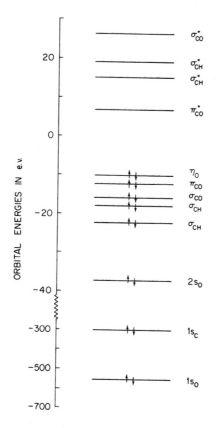

Figure 2   Energies and bonding character for the
molecular orbitals of formaldehyde.

Although the energies and bonding character of each MO depends upon the particular molecule involved, some generalizations can be made regarding the general pattern of orbitals in most polyatomic organic molecules; a symbolic representation of this pattern is shown below.

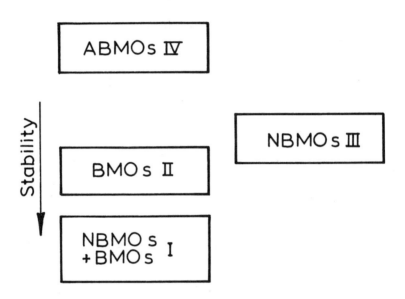

Here area I represents electrons held very tightly to the molecule, whether in bonding or nonbonding MOs. The orbitals here would include inner-shell types, "s" lone pair electrons of electronegative elements such as 0 and F, etc. The next area, II, includes electrons participating strongly in both σ and π bonding. Note that all MOs associated with both areas I and II are completely filled in the ground state of the molecule (as indicated by shading) whereas the antibonding orbitals (area IV) complementary to the BMOs in area II are all empty (no shading) in the ground state. If the molecule contains "p" type lone pairs of electrons (e.g. as in N,0), there will be one or more doubly-occupied NBMOs (area III) in addition.

According to molecular orbital theory, the low-lying excited states of such molecules correspond to electron configurations in which an electron in one of the least stable BMOs in area II (or in an NBMO) is promoted to one of the more stable antibonding MOs (in area IV). Thus the photon energy necessary to induce the lowest-energy excitation is directly proportional to the distance ΔE between the top of areas II and III and the bottom of the ABMO area IV. In most organic compounds containing no multiple bonds, the distance ΔE between the highest occupied BMO and the lowest unoccupied ABMO is very large (> 100 kcal/mole), and no transitions are seen in the near ultraviolet and visible wavelength region. For compounds containing double and triple bonds, however, the energy gap is much reduced since the BMOs associated with the π electrons are not as stabilized as those for σ electrons, and the ABMOs for π electrons are not as destabilized as those for σ electrons. Thus the excitation of an electron from a π to a π^* orbital requires much less energy than a corresponding $\sigma \rightarrow \sigma^*$ excitation. A rough comparison between the levels in n-butane and trans-butadiene is given by the following symbolic diagram:

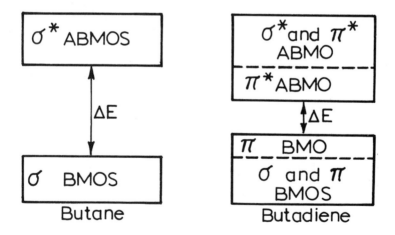

Experimental confirmation of the relative energies and nature of the higher occupied MOs in such systems is possible from the measurement of ionization potentials, since the minimum energy required to remove an electron from a molecule should correspond to the energy of the least stable level. As predicted from the diagrams above, the first ionization potential for trans-butadiene of 9.0 e.V. does correspond to removal of a π electron, and the energy required is substantially greater than that of 10.6 e.V. found for n-butane[5].

In the case of conjugated hydrocarbons - both acyclic polyenes and cyclic systems, most (if not all) of the transitions observed in the near ultraviolet and visible spectrum are assigned to transitions of an electron from a π-bonding MO to a π^*-antibonding MO. The exact photon energy required depends upon the number of double bonds conjugated together and upon the manner in which this conjugation occurs. For linear polyenes, the ΔE for the lowest-energy $\pi \rightarrow \pi^*$ transition decreases with chain length, a phenomenon which is readily rationalized by theory if the assumption is made that each π electron travels freely along the entire length of the conjugated chain. The π electron energy levels associated with such an electron should be roughly comparable to those for a particle travelling in a one-dimensional path along which the potential is constant and infinite , but which becomes infinitely large outside the chain ends. The quantum-mechanical prediction of the allowed orbital energies ϵ for such π electrons is given by

$$\epsilon = \frac{n^2 h^2}{8mL^2}$$

where h is Planck's constant, m the electron mass, L the length of the one-dimensional "box" and n is a quantum number with allowed values n = 1, 2, 3, --- ∞. If the average

---

5.   D.W. Turner in <u>Adv. Phys. Org. Chem.</u>, <u>4</u>, 31 (1966).

distance between each adjacent pair of carbon atoms is assumed to be some constant value X, then a polyene containing M conjugated double bonds will have length

$$L = (2M-1)X$$

The ground state of such a system has all π orbitals doubly occupied from $n = 1$ to $n = M$, so the lowest energy $\pi \rightarrow \pi^*$ transition should correspond to the energy required to excite an electron from $n = M$ to $n = M + 1$; from the equation above

$$\Delta E = \frac{(M+1)^2 - M^2}{(2M-1)^2} \frac{h^2}{8 \, mX^2} = \frac{2M+1}{(2M-1)^2} \frac{h^2}{8 \, mX^2}$$

$$\text{or} \quad \Delta E = K \frac{2M+1}{(2M-1)^2}$$

where K is a constant independent of chain length. For long chains, $M \gg 1$, and the transition energy becomes inversely proportional to chain length since

$$\Delta E \sim \frac{K}{2M}$$

Since ΔE is inversely proportional to the photon wavelength, this model predicts that

$$\lambda_{max} \propto (2M-1)^2/(2M+1)$$

or in the limit of long chains

$$\lambda_{max} \propto 2 M$$

A comparison of the predicted versus the experimental dependence of λ_{max} for the S_o - S_1 $\pi \rightarrow \pi^*$ transition upon chain length is illustrated in Figure 3; note that the relationship is semiquantitative except for very short chains. Introduction of electron-electron repulsion effects

41

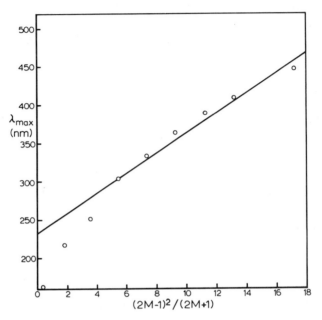

Figure 3   Correlation λ_{max} with M for polyenes
S_o - S_1 transition

are required to improve the accuracy of the correlation (<u>vide</u> <u>infra</u>).

Thus far only $\pi\pi^*$ transitions have been considered. As indicated previously, however, transitions arising from nonbonded electrons to π^* orbitals are also possible in those compounds containing nitrogen and oxygen which possess "p" lone pairs of electrons. Such " $n\pi^*$ " transitions often occur in the UV and visible range, with the actual transition energy again depending upon the number of double bonds, etc. Consider the example of the series of chains which contain a sequence of conjugated double bonds which terminate with a -C = 0 rather than a -C = CH_2 group. For the first member of the series, formaldehyde, the NBMO is much less stable than the π bonding MO and the longest wavelength transitions are nπ* rather than $\pi\pi$*. As increasing numbers of C = C bonds are introduced into the molecule, the lowest π* MO becomes progressively more **stable** whereas the NBMO position remains the same, so that the ΔE for the $\pi \rightarrow \pi$* transition will require a lower transition energy than that required for nπ*, and thus the character of the longest wavelength transition will be altered from nπ* to $\pi\pi$*. The three cases of interest are indicated in the diagrams below:

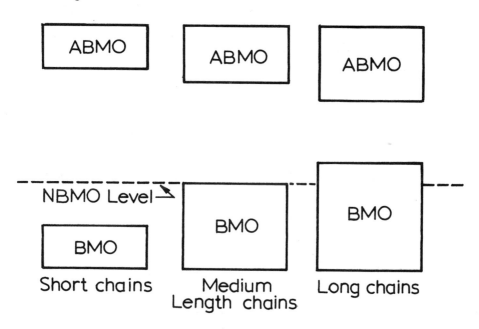

Consider as examples the <u>triplet</u> states resulting from the n $\rightarrow$ π* transitions in the following types of carbonyl compounds:

The only case here which falls into the " short chain " category is I (formaldehyde, etc.), since the transition to the $^3n\pi^*$ state at ~ 400nm requires much less energy than that to the $^3\pi\pi^*$ state (estimated at ~ 250 nm from theoretical calculations).[6]   Although conjugation of one C = C bond to give II does lower the π^* MO energy and the $^3n\pi^*$ state of acrolein and derivatives occurs at lower energy of ~ 420 nm, the highest-occupied π MO in II is much less stable than in I, with the result that the $^3\pi\pi^*$ state occurs at almost the same wavelength[7] - thus the " medium length chain " case is achieved with one C = C conjugated to one C = O. The " long chain " case is met for all type III systems, since the energy required to reach the $^3n\pi^*$ state decreases only slowly with increasing conjugation, whereas the $^3\pi\pi^*$ state transition energy decreases much faster. For example, the 0 - 0 bands for m = 2,3,4 occur at $\sim$660, $\sim$790 and $\sim$900 nm respectively[8].

Although only conjugated chains have been dealt with thus far, the same type of simple considerations can be applied to cyclic systems. The π MOs of benzene, for example, can be considered as those appropriate to particles travelling around a ring of constant potential. The accessible energy levels for such a particle are

$$\epsilon = \frac{n^2h^2}{8\pi^2mr^2}$$

where r is the radius of the ring, and the allowed values for the quantum number n are

$$n = 0, \pm 1, \pm 2, \text{--------} \pm \infty$$

6.   R.J. Buenker and S.D. Peyerimhoff, J. Chem. Phys. 53, 1368 (1970).

7.   D.R. Kearns, G. Marsh and K. Schaffner, J. Chem. Phys. 49, 3316 (1968).

8 .   D.F. Evans, J. Chem. Soc., 1735 (1960).

Note that the most stable level (n = 0) is nondegenerate, whereas all higher levels occur in degenerate pairs (since n = +1 yields the same energy as n = -1, etc.). If it is assumed that the radius r is proportional to the number of double bonds, p, around the periphery of a ring, then the longest wavelength absorption λ of the hydrocarbon should depend on p as

$$\lambda \propto \frac{p^2}{2p+1}$$

or, for large p

$$\lambda \propto p$$

An example of the trend predicted above is illustrated in Figure 4, in which the experimental $\lambda_{(0-0)}$ for the lowest S_0 - T_1 transition in linear polyacenes[9] is plotted against $p^2/2p+1$. The correlation is quite good, considering the crudeness of the model and that a polyacene with cross links has been approximated as a circle without cross links.

Although the gross properties and trends in $\lambda_{absorption}$ for unsaturated organic molecules can be understood using the " independent particle " techniques described above, neither these theories, nor other methods such as simple Hückel calculations, can account for some of the more subtle features of excited state energies. For example, all such theories predict that the energy of an excited state should be independent of the spin orientation of the electrons in the singly-occupied MOs, whereas the energy difference between the singlet and triplet states for the same

9.   S.P. McGlynn, T. Azumi and M. Kinoshita, "Molecular Spectroscopy of the Triplet State", Prentice-Hall, Inc., Englewood Cliffs, New Jersey, 1969, Chapter 3.

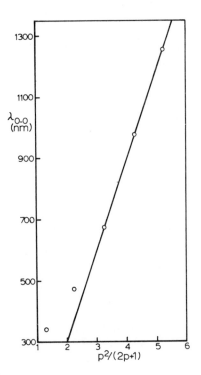

Figure 4   Correlation λ_{0-0} with p for polyacene

                      S_0 - T_1 transition

electron configuration amounts to 30-70 kcal/mole in the polyenes. In addition, the simple theories become useless in considering the excited states of some diatomic and triatomic molecules. For example, independent particle theories cannot account for the energy gap between the ground state and the two lowest lying excited states of the O_2 molecule, all of which have the same basic electron configuration $(\pi)^4(\pi^*)^2$. That these states do, in fact, possess both different energies and different equilibrium

46

bond distances has been determined experimentally[10]. In order to explain the manner in which the same electron configuration can yield two or more states of different energy, the role of electron-electron repulsions must be investigated.

The Role of Electron Repulsion

In principle, the calculation of the total energy of repulsion between all the electrons in a molecule is a simple matter – all we need do is to consider each pair of electrons, i and j, and to compute the time-average electrostatic repulsion between them

$$E_{repulsion} = \sum_i {}_{<} \sum_j \frac{e^2}{(r_{ij})_{avg}}$$

In practice, however, it is impossible to evaluate accurately the time-average distance, $(r_{ij})_{avg}$, between each pair! Given the orbital electron configuration for a state <u>and</u> the restrictions imposed by the Pauli Exclusion Principle with regard to the overall wave function, it is possible nevertheless to understand the reason that the same configuration can yield states of greatly different energy.

Consider a molecular electronic configuration A^1B^1 in which two MOs, A and B, each contain one electron and in which all other MOs are either doubly-filled or empty. With respect to A and B, one could write a total wave function

$$\psi = A(1)B(2)$$

in which some electron labelled #1 occupies A and an electron labelled #2 occupies B. Unfortunately this wave function implies that we can physically distinguish between the two electrons when, in fact, we cannot. In other words, the wave function A(2)B(1) is just as likely as the one written above. To introduce the concept of physical indistinguishability between electrons, we must insist that ψ^2 be symmetrical with respect to an **interchange** of the electron labels.

---

10. G. Herzberg, "Spectra of Diatomic Molecules", D. Van Nostrand Company Inc., Princeton, New Jersey, 1950, p. 446. See also the discussion in Chapter 3.

The only proper wave functions which fulfill this requirement are

$$\psi^+ = \frac{1}{2}[A(1)B(2) + A(2)B(1)]$$

and

$$\psi^- = \frac{1}{2}[A(1)B(2) - A(2)B(1)]$$

Note that ψ^+ itself is <u>symmetric</u> with respect to electron
interchange, whereas ψ^- is <u>antisymmetric</u> (i.e. it changes
sign if electron 1 is replaced by electron 2, and vice
versa). If electron-electron repulsion is not included in
the calculation of the total energy, ψ^+ and ψ^- yield states
of equal energy. However, when electron repulsion is included properly, the degeneracy is broken because the
electron repulsion between electrons 1 and 2 for ψ^+ is
different from that computed for ψ^-. This can be proven by
substituting the wave functions above into the formula for
the " expectation value " of the repulsion energy

$$E_{repulsion} = \int \psi [\frac{e^2}{r_{12}}] \psi dV$$

Substitution of ψ^+ for ψ leads to

$$E^+_{repulsion} = J_{AB} + K_{AB}$$

$$E^-_{repulsion} = J_{AB} - K_{AB}$$

where $J_{AB} = \int\int A(1)^2[\frac{e^2}{r_{12}}]B(2)^2 d\tau(1)d\tau(2) =$
$\int\int A(2)^2[\frac{e^2}{r_{12}}]B(1)^2 d\tau(1)d\tau(2)$

and K_{AB}, the " exchange integral " is given by

$$K_{AB} = \int\int A(1)A(2)[\frac{e^2}{r_{12}}]B(1)B(2)d\tau(1)d\tau(2)$$

Since $E^+_{repulsion} > E^-_{repulsion}$, the average value of r_{12} for
the wave function ψ^+ must be smaller than that for ψ^-. For
any state described by a symmetrical wave function ψ^+, the
electrons are closer together on the average (and thus the

48

net repulsion larger) than for an antisymmetric wave function ψ^-.

Thus far, we have established that a given electron configuration A^1B^1 gives rise to two types of states of different energy. To identify these with singlets and triplets, it is necessary to specify the spin wave function. The spin functions we are allowed to use are constrained by the Pauli Exclusion Principle which states that the total wave function for any state, after the spin part has been added, must be antisymmetric with respect to electron interchange. Since ψ^+ is symmetric, its spin function must be antisymmetric, and since ψ^- is already antisymmetric, its spin function must be symmetric. If the electron spins are labelled α and β, the only proper antisymmetric spin function for A^1B^1 is

$$[\alpha(1)\beta(2) \; - \; \alpha(2)\beta(1)]/\sqrt{2}$$

whereas three symmetric spin functions are possible

$$\alpha(1)\alpha(2)$$
$$\beta(1)\beta(2)$$
$$\text{and} \quad [\alpha(1)\beta(2) \; + \; \alpha(2)\beta(1)]/\sqrt{2}$$

The antisymmetric spin function yields a net spin $S = 0$ for the molecule and thus we identify the total wave function

$$\psi^+[\alpha(1)\beta(2) \; - \; \alpha(2)\beta(1)]/\sqrt{2}$$

with the singlet state associated with A^1B^1, whereas each of the three symmetric spin functions yields a net electron spin of $S = 1$; thus

$$\psi^- \; [\alpha(1)\alpha(2)]$$
$$\psi^- \; [\beta(1)\beta(2)]$$
$$\text{and} \quad \psi^- \; [\alpha(1)\beta(2) \; + \; \alpha(2)\beta(1)]/\sqrt{2}$$

are the three components of the triplet wave function. Since it has been established above that the total electron repulsion for the ψ^+ wave function is worse than for ψ^-, thus <u>the triplet of any A^1B^1 configuration must be more stable than the singlet</u>. Note that this " split " in energy between the singlet and triplet is a direct consequence of the fact that electrons are indistinguishable, and the resulting wave functions which are allowed simply have different time-average electrostatic repulsions. Electron spin, per se, has nothing to do with this energy splitting except to help us identify the total spin quantum number associated with ψ^+ and ψ^-. In other words, "electrons of like spin tend to keep away from each other " only because the spatial wave function for a triplet must be antisymmetric, and this type of wave function yields a smaller net energy of repulsion than does the spatially-symmetric one.

Some idea of the energy difference between the singlet and triplet can be deduced if the form of the MOs A and B are known. Note that the integral K_{AB} will be large only if A and B overlap well in space, since if they do <u>not</u>, the products $A(1)B(1)$ and $A(2)B(2)$ will be small everywhere and $K_{AB} \sim 0$. If A represents a bonding π MO and B an anti-bonding π MO, the product $A(1)B(1)$ is substantial since the MOs overlap to a great extent; thus K_{AB} is large and a substantial singlet-triplet split is both expected and indeed observed (30-70 kcal/mole in polyenes, for example). In contrast, if A is an NBMO which is orthogonal to the π system (e.g. the lone pair NBMO in formaldehyde) and B is an π ABMO, then $A(1)B(1)$ is small in most regions of space and the singlet-triplet split is both expected and found to be quite small (typically a few kcal/mole for the $^3n\pi^*$ state of formaldehyde). Figure 5 may be helpful in remembering these trends; some typical S_1 - T_1 splitting values for organic molecules are given in Table 1 **(see Ref. 9)**.

TABLE 1

Molecule	Excitation Type	$S_1 - T_1$ Splitting in kcal/mole
Ethylene	$\pi\pi^*$	69
Benzene	$\pi\pi^*$	53
Naphthalene	$\pi\pi^*$	39
Formaldehyde	$n\pi^*$	9
Acrolein	$n\pi^*$	5
Glyoxal	$n\pi^*$	7

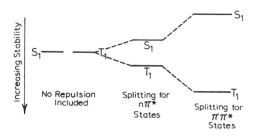

Figure 5   Singlet-triplet separation for $n\pi^*$ and $\pi\pi^*$ states.

Although all the considerations given above are in molecular orbital terms, a rough translation into valence-bond diagrams is possible. Consider the $\pi\pi^*$ states of ethylene as an example. In the <u>triplet</u>, the two unpaired π electrons tend to keep " away from each other " since $(r_{12})_{avg}$ is large; quantitative calculations predict that there is almost a 0% chance of finding both electrons simultaneously on the same carbon atom.   Thus the $^3\pi\pi^*$ state is best represented as a " biradical "

$$\ce{H\ H >C-C< H\ H}$$

In contrast, there is almost a 100% chance of simultaneously finding the two π electrons on the same carbon atom in the $\pi\pi^*$ singlet, so that the best valence-bond diagram is a hybrid of the two structures.

$$\ce{H\ + - H >C-C< H\ H} \longleftrightarrow \ce{H\ - + H >C-C< H\ H}$$

In the case of $n\pi^*$ states, both the singlet and the triplet act approximately as biradicals, so formaldehyde is best represented in both states as follows, with a " three electron " pi bond:

$$\ce{H\ >C-O. H}$$

The Geometry of Excited States

It is commonly accepted that the characteristic structure of unsaturated organic molecules in their <u>ground</u> states - both in terms of bond lengths and bond angles - is due to rather stringent geometrical requirements for maximization of the π bonding energy. Since electronic excitation in such systems usually involves a substantial change in the π bonding, it should come as no surprise that the optimum geometry in $\pi\pi^*$ and $n\pi^*$ states usually differs considerably from that in the ground state.

A few qualitative conclusions concerning the structure of excited states can be deduced from simple valence-bond structures. For example, the considerations in the previous section indicated that the $^3\pi\pi^*$ state of ethylene is well represented by the biradical structure $\dot{C} - \dot{C}$, from which the following implications can be drawn:

i)   the equilibrium carbon-carbon bond length in the trip-let should be closer to that for the ground state of ethane than for ethylene, since excitation has completely destroyed the π bond, thus reducing the carbon-carbon bond order from 2 to 1;   ii)   the large activation energy necessary to isomerize the carbon-carbon double bond in the ground state should be absent in the triplet, again because the multiple bonding has been destroyed.

Similar trends can be predicted for the lowest $^3\pi\pi^*$ state of butadiene, represented as a resonance hybrid of three biradical structures

$$C = C - \dot{C} - \dot{C} \longleftrightarrow \dot{C} - C = C - \dot{C} \longleftrightarrow \dot{C} - \dot{C} - C = C$$

Thus the 1-2 and 3-4 bonds should be <u>longer</u> in the triplet than in the ground state but the 2-3 bond should become shorter, and similarly the barrier to rotation about the terminal bonds should decrease whereas that about the cen-tral bond should increase.

Molecular orbital calculations indicate that semi-quantitiative predictions concerning the structure of exci-ted states can be made by considering the changes in energy with geometry of the least stable bonding π MO, and the most stable antibonding MO in such systems, since the <u>total</u> bonding energy of the other occupied MOs is rather insensi-tive to geometry[11].

Consider an acyclic polyene XY which is twisted about one bond, r-s which is a double bond in the ground state,

---

11.   N.C. Baird and R.M. West, <u>J. Amer. Chem. Soc.</u>, 93, 4427 (1971).

for example, the 3-4 bond in hexatriene.

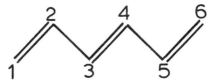

The π MOs of twisted XY do not extend along the entire molecule but are associated with one or the other of the orthogonal π networks X and Y which meet at the r-s bond. Since a <u>double</u> bond is twisted, both X and Y possess odd numbers of unsaturated carbon atoms and the π MOs of X and Y are those for the free radicals (XH)$\cdot$ and (YH)$\cdot$. In particular, the two unpaired electrons must occupy " non-bonded " MOs which shall be denoted as ϕ_X and ϕ_Y. If XY becomes planar, these two NBMOs interact and combine to form one bonding MO ψ^+ and one antibonding MO ψ^- (see diagram below)

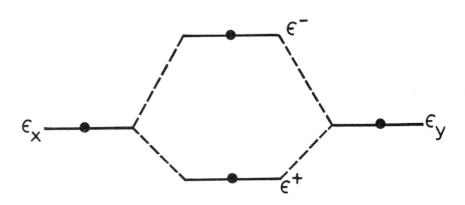

The essential point here is that <u>the bonding MO ψ^+ is not</u> as stabilized (relative to ϵ_X and ϵ_Y) <u>as the antibonding MO ψ^- is destabilized</u>. Thus, if both ψ^+ and ψ^- are each populated by one electron, as in a $\pi\pi^*$ excited state, the <u>net</u> effect of the interaction in the planar species involves a

<u>loss</u> rather than a gain in π energy! In summary, <u>MO theory predicts that in $\pi\pi^*$ states of polyenes, the optimum molecular geometry has one C=C bond twisted by 90°!</u> In effect, excitation of one π electron from a bonding to an antibonding MO alters not only the magnitude of the barrier to rotation about a C=C bond but even the sign of the barrier. A comparison of the potential curves for the ground and lowest triplet of polyenes is shown below:

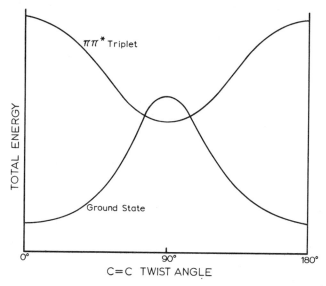

The magnitude of the energy difference between the planar and twisted triplet is predicted to be proportional to the product of the spin densities for the unpaired electrons at the position (r-s) of twisting; thus the barrier to planarity should be greatest for ethylene

$$\overset{\uparrow}{C} - \overset{\uparrow}{C}$$

but much smaller for butadiene and more conjugated polyenes, for which the spin density is more delocalized:

$$C = C - \overset{\uparrow}{\underset{r}{C}} - \overset{\uparrow}{\underset{s}{C}} \longleftrightarrow \overset{\uparrow}{C} - C = C - \overset{\uparrow}{\underset{r}{\underset{\;}{C}}}_{\;\;s} \longleftrightarrow \overset{\uparrow}{C} - \overset{\uparrow}{C} - \underset{r}{C} = \underset{s}{C}$$

One other interesting prediction emerges from the potential energy diagram above - namely that the 90° twisted triplet polyene is slightly more stable than the twisted **singlet ground state!** The energy difference between S_O and T_1 in this conformation amounts to only a kcal/mole or so, in contrast to the huge splitting (40 - 100 kcal/mole) when the twist angle is zero. Intersystem crossing from T_1 to S_O should be greatly facilitated at the crossing point (very close to 90°) and thus any triplet which can twist in this way should have a very short lifetime!

It should be noted that the potential curves discussed above are valid only for $\pi\pi^*$ and not for $n\pi^*$ excited states. In the latter, the bonding MO ψ^+ is occupied by two electrons and the antibonding MO ψ^- by only one, so that the planar geometry should still be preferred to the twisted for $n\pi^*$ states. The net barrier to rotation about a $C = C$ bond will, nevertheless, be somewhat reduced in $n\pi^*$ states as compared to that in the ground state.

One important factor which may alter the rotation barriers in both $\pi\pi^*$ and $n\pi^*$ states but which is neglected in the arguments above is the influence of hyperconjugation upon orbital energies. Hyperconjugation refers to the interaction of π orbitals with MOs which are involved in forming single C-H and C-C bonds. For example, in twisted ethylene the p_π orbital of atom r is not forbidden by symmetry to interact with some of the orbitals which form the C-H bonds at atom s, and vice versa. Although it is usually assumed that such hyperconjugation stabilizes a twisted molecule, the extent to which he rotation barrier is altered thereby is still not well-known.

The discussion above can be summarized by stating that the unpaired electrons in $\pi\pi^*$ triplets interact in a destructive rather than a constructive manner, and that the molecule takes steps to minimize this interaction. Although the mechanism of reducing the destabilization is obvious for acyclic polyenes, what course of action is available to molecules which cannot twist due to torsional strain effects (e.g. cyclic olefins with small-membered rings and aromatic

hydrocarbons) or in which twisting is a meaningless process (e.g. linear alkynes)? Although the molecular orbital
theory for the $^3\pi\pi^*$ states of such systems is still under
development, several obvious possibilities are open to any
molecule which wishes to reduce interaction between the unpaired electrons. One possibility is to lengthen the
bond(s) which join the two biradical components. This
would imply that, for example, the D_{6h} symmetry of the
ground state of benzene is broken in the lowest triplet in
favour of either a quinoidal or/and a biallylic structure:

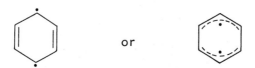

In the case of acetylene, the destructive interaction could
be decreased, again by reducing the molecular symmetry, to
yield a nonlinear triplet:

For small molecules, some progress in deducing changes
in shape upon electronic excitation can be made by plotting
orbital energies versus bond angles for individual molecular types.

Such <u>Walsh Diagrams</u>[12] can be deduced and used rather
simply for small polyatomics such as formaldehyde and acetylene. As an example, the Walsh Diagram which follows the
effect upon the valence-electron orbital energies of vary
the H-C-C angles from 90° to 180° in acetylene is given in
Figure 6. Since the number of valence electrons in C_2H_2 is

12.   A.D. Walsh, <u>J. Chem. Soc.</u>, 2260, 2266, 2288, 2296,
       2301, 2306 (1953); see also C.J. Ballhausen and
       H.B. Gray, "Molecular Orbital Theory", W.A. Ben-
       jamin, Inc., New York, 1964.

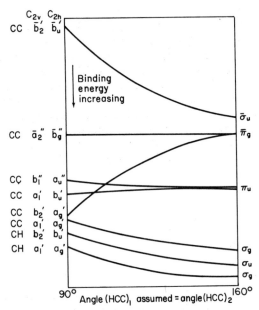

Figure 6   Walsh diagram for acetylene.
Reprinted with permission from J. Chem. Soc.,
2288 (1953).

10, all orbitals up to and including the doubly-degenerate π_v are doubly-occupied in the ground state. Inspection of the curvature of the energy plots indicates that acetylene should be linear in the ground state (as observed experimentally) since the occupied σ orbitals are most stable for $\angle HCC=180°$ and the π_v orbital energies are essentially independent of angle. However, occupation by an excited electron of the π_g orbital (as in the lowest $\pi\pi^*$ excited state) should yield a <u>nonlinear</u> molecule, since the energy of one of the degenerate (at 180°) π_g MOs decreases sharply as the HCC angles are decreased. Similarly, the Walsh Diagram for formaldehyde predicts a loss of polarity in the $n\pi^*$ excited states. Both experiment and <u>ab initio</u> calculations confirm these predicted geometry changes upon elec-

tronic excitation for acetylene[12, 13] and for formaldehyde[12, 14]. One would anticipate that MO calculations of
this type will be extended to study intramolecular potential surfaces of even larger excited molecules in the next
few years; indeed an extensive set of calculations for
acrolein has already been executed[15].

Photochemical Reactions

The semiquantitative MO theory for polyenes described
in the previous section and supported by explicit MO calculations of various types[16 - 18] indicates a preference
for a 90° twist of one C=C bond in the lowest $\pi\pi^*$ triplet
state. Since the conformational integrity about this bond
is lost by such a rotation, MO theory predicts that cis-
trans isomerization should occur readily in such states.
(See also the discussion in Chapter 7 .) The MO calculations also predict an energetic preference for internal,
rather than terminal, C=C bond isomerization in long conjugated polyenes[16]. Complementary to the prediction that
planar conjugated polyenes should become twisted in the
excited state is the prediction that allene and other C_nH_4
cumulenes, which are _twisted_ in the _ground_ state, should

13.   W.E. Kammer, _Chem. Phys. Letters,_ 6, 529 (1970).

14.   R.J. Buenker and S.D. Peyerimhoff, _J. Chem. Phys.,_
       53, 1368 (1970).

15.   A. Devaquet and L. Salem, _Can. J. Chem.,_ 49,
       977 (1971).

16.   N.C. Baird and R.M. West, _J. Amer. Chem. Soc.,_
       93, 4427 (1971).

prefer a <u>planar</u> conformation in the lowest $\pi\pi^*$ states[17 - 19] - thus cis-trans isomerization should occur upon excitation of substituted cumulenes.

In recent years, substantial progress has been achieved in predicting and rationalizing both intramolecular and intermolecular reactions using qualitative molecular orbital theory. In particular, the formalism developed by Woodward and Hoffman[20] allows one to decide whether a given reaction should be kinetically fast process due to the small activation energy required ("allowed reaction") or slow due to a large activation energy ("forbidden reaction"). The strategy adopted by Woodward and Hoffman is to relate the bonding and antibonding character of the orbitals of the reactant to those of the product by use of symmetry arguments and by relationships based upon simple MO theory. If the occupied molecular orbitals of the reactants are identical in character to those of the product, then the reaction is allowed. In contrast, if the orbital electron configuration of the reactants differs from that of the products, the reaction should be forbidden. The most important **result** of these correlations for photochemistry is that, for <u>almost</u> all concerted reactions, the photochemical reaction path is allowed if the thermal path is forbidden, and vice versa!

As an explicit example of the use of such orbital correlation diagrams, consider the conversion of cis-butadiene to cyclobutene:

17.   R. Hoffman, <u>Tetrahedron</u>, <u>22</u>, 521 (1966).

18.   Y.A. Kruglyok and A.G. Dyadyusha, <u>Theoret. Chim. Acta</u>, <u>10</u>, 23 (1968).

19.   L.J. Schaad, L.A. Burnelle and K.P. Dressler, <u>Theoret. Chim. Acta</u>, <u>15</u>, 91 (1969).

20.   R.B. Woodward and R. Hoffman, "The Conservation of Orbital Symmetry", Academic Press, New York, 1970.

The orbitals of the product are quite similar to those of the reactant in terms of both their bonding character and their energy, with the exception of two MOs - those corresponding to the least stable π bonding MO (π_2) and the most stable π^* antibonding MO (π_3) of butadiene. As mentioned previously, any particular MO is bonding with respect to some interactions and antibonding with respect to others. For butadiene, π_2 is <u>net</u> bonding but <u>antibonding</u> with respect to the C_1 - C_4 interaction; in contrast π_3 is net antibonding but bonding with respect to the C_1 - C_4 interaction. During the disrotatory closure of butadiene, the C_1 - C_4 antibonding effect in π_3 is greatly increased in magnitude since the π orbitals on the end carbons are twisted toward each other. This is illustrated in Figures 7a and 7b. The net effect is so strong that this MO which was a net bonding π MO in the reactant is converted to net antibonding σ^* MO in the product. The reaction also affects the antibonding π^* MO (π) of butadiene but in a <u>stabilizing</u> sense, since the C_1 - C_4 interaction is bonding in π_3 and this bonding component eventually dominates over its antibonding component in the product molecule. These orbital correlations are summarized in the figure below.

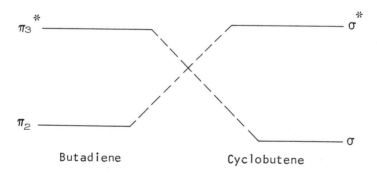

Butadiene   Cyclobutene

61

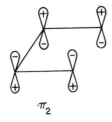 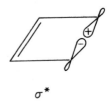

π_2 σ^*

Figure 7a

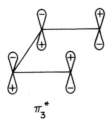 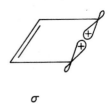

π_3^* σ

Figure 7b

Thus the overall reaction requires a large activation energy if one starts from the ground-state $\pi_2{}^2\pi_3{}^0$ electron configuration, but little or no energy if one starts from the excited state $\pi_2{}^1\pi_3{}^1$ configuration (since the energy loss by the electron in π_2 is compensated by an energy gain in π_3). Thus the ground state disrotatory closure of butadiene is forbidden but the same reaction is allowed if one starts with the first excited state of butadiene!

Since the bonding characteristics for many systems can be deduced from simple MO theories, the orbital correlation analysis can, and has been, applied to a great variety of organic photochemical reactions. The reader is referred to the excellent review article available[21] for further applications.

Before concluding these remarks on the theoretical analysis of photochemical reactions, a word of caution should be introduced. In determining whether a reaction involving ground states of molecules is feasible or not, the activation energy criterion is usually sufficient. In photochemical reactions, however, a small activation energy does not always ensure that the reaction will proceed quickly, since the spin quantum numbers of the products must correlate with those of the reactants. That is, a reaction may proceed only slowly or not at all even though the individual orbitals of reactants correlate properly with those of the product, if the total spin angular momentum of the products is not equal to that for the reactants. The effect of this requirement is much more pronounced for reactions involving small molecules than for organic systems; the reader is referred to the review by Evleth for a useful discussion of this topic[22].

---

21.   R.B. Woodward and R. Hoffman, op. cit.

22.   E.M. Evleth, "Photochemistry of Macromolecules", Plenum Press, 1970.

CHAPTER 3
ELECTRONIC SPECTRA
OF EXCITED STATES

Introduction

Light absorption by molecules is structured into regions, usually known as <u>bands</u>, which mark the different excitation processes from the ordinary ground state to electronically-excited states of the molecule. Each band of the spectrum has, in principle, a sub-structure of vibrational quantum changes and a sub-sub-structure of rotational quantum changes, though these details are fully developed only in vapor phase spectra, that is, under conditions where individual molecules are relatively isolated from one another. The separate bands are characterized by intensity and position or, put another way, by the probability of the electronic transition and by its location on a wavelength or frequency (energy) scale. It is convenient to describe the states connected by a transition in terms of the <u>electron configurations</u> introduced in Chapter 1 which indicate the address of the outer electrons before and after the transition. In the familiar example of an $n \rightarrow \pi^*$ transition of a simple saturated carbonyl compound the electron configurations of the initial and final states are, respectively,

$$\ldots \ldots (\pi_{CO}, \ b_1)^2 (n_0, \ b_2)^2 (\pi^*_{CO}, \ b_1)^0 \ : \ ^1A_1 \qquad (1)$$

$$\ldots \ldots (\pi_{CO}, \ b_1)^2 (n_0, \ b_2)^1 (\pi^*_{CO}, \ b_1)^1 \ : \ ^{1,3}A_2 \qquad (2)$$

so that the transition promotes one electron only, the remainder occupying the same orbitals as in the electronic

ground state. Since sensibly all electron transitions are
one-electron promotions, the abbreviated description as
$n \rightarrow \pi^*$ - the one-electron feature being understood, but not
stated - is perfectly serviceable. It should be realized,
however, that the configuration (2) corresponds not to one
state but to two, since the spins of the electrons occupying the unfilled orbitals may be either parallel (the
'triplet' state) or antiparallel (the 'singlet' state):
this is conveyed by the superscripts (1 = singlet, 3 =
triplet) attached to the group theoretical symbol A_2 on the
right side of Equation (2). In order to incorporate this
information into the shorthand notation it will be necessary to specify the electronic $n \rightarrow \pi^*$ transition as singlet-
triplet or singlet-singlet. For the reasons outlined in
Chapter 2, a triplet state lies lower in energy than the
singlet state of the same electronic configuration. The
group-theoretical symbols A_1 and A_2 are discussed later.

In order that a transition may occur by light absorption there must be some form of interaction between the
radiation and the molecule. With few exceptions the radiation acts through the action of its electric field upon an
electric dipole moment in the molecule. The greater the
electric moment change accompanying the transition the
greater the absorption intensity (strictly, the intensity
is proportional to the square of the moment change); thus
one reason underlying differences in intensity among the
various bands, probably the most obvious feature of any
spectrum, is a variation in the dipole moment change. In
some transitions the electric moment change is small, or
zero, and the transition may then pass unseen in the spectrum.

The mechanism just described is not the only means of
interaction between light and molecules. In NMR and ESR
transitions take place through the action of the magnetic
field of the radiation upon a magnetic dipole, namely that
associated with nuclear or electron spins. These magnetic
-dipole transitions are however inherently weaker than the
electric-dipole transitions and so are unimportant unless
the electric-dipole mechanism is forbidden. In electronic
spectroscopy, an allowed electric-dipole transition has

about 10^5 times the intensity of an allowed magnetic-dipole transition.

Intensity, in this context, should be understood to mean the integrated intensity associated with an electronic band, that is, the total area under an absorption envelope plotted as a curve of molar absorptivity (ϵ, ℓ mole^{-1} cm^{-1}) against wave number ($\bar{\nu}$, cm^{-1}). Its measure is the <u>oscillator strength</u> f, which is the ratio of the experimental transition probability to an ideal value,

$$\int \epsilon d\bar{\nu} = \frac{\pi e^2 N_o}{2303 \; mc^2} = 2.31 \times 10^8 \tag{3}$$

Here, N_g is Avogadro's number, m and e are the electron mass and charge, and c is the velocity of light. The reciprocal of 2.31×10^8 is 4.33×10^{-9}, thus

$$f = 4.33 \times 10^{-9} \int \epsilon d\bar{\nu} \tag{4}$$

An allowed electric-dipole transition has a f-value of the order unity. The f-value is a more informative quality than ϵ_{max} because different bands occupy frequency regions of different breadth: according to the Franck-Condon principle (see Chapter 1), the greater the structural change between the ground and excited states of a transition the greater the frequency range enclosed by the band. A rough estimate of the f-value can be made from the formula,

$$f = 4.33 \times 10^{-9} \times a \times \epsilon_{max} \times \Delta\bar{\nu}_{\frac{1}{2}} \tag{5}$$

in which ϵ_{max} is the peak molar absorptivity, and $\Delta\bar{\nu}_{\frac{1}{2}}$ is the full width of the band at one-half the peak height. The constant $a = (\pi/2.76) = 1.07$ if the band envelope is gaussian, that is, shaped like a normal error curve, but in fact bands are seldom symmetrical and for qualitative purposes, it is sufficient to put $a = 1$.

Singlet-triplet transitions in organic molecules are always very weak ($f < 10^{-6}$). A transition connecting pure

singlet and triplet states is totally forbidden in absence
of spin-orbit coupling (see p. 76, this chapter), which,
however, mixes minor amounts of singlet character into the
triplet wave function, and vice versa, so that the transitions become allowed to the extent that such mixing enters.

Classification of Electronic States

The orbital symmetry of an electronic state is an important characteristic of that state and can be used, in
conjunction with some simple rules, to determine the conditions under which a transition from the ground state to
the state in question may occur by absorption of light
quanta. These rules are explained in this and the two
following sections. (Readers whose interest lies more with
the results than with the theory may proceed directly to
the discussion of actual spectra on p. 76 .)

An electronic state is characterized by the symmetry
of its orbital wave function and by the multiplicity of its
electron spin function. With few exceptions, stable molecules in their ground state have completely filled orbitals
so that the spin function has unit multiplicity (singlet
states). Thus the ground state of, for example, CH_2O with
all electrons spin-paired [see Equation (1)] is written $^1A_1$
("singlet A one") in which the prefix gives the multiplicity
and the remainder of the symbol the group-theoretical
species classification of the orbital wave function.

The symmetry classification of wave functions is based
on the symmetry properties of molecules. Most small molecules possess certain symmetry elements such as a plane (σ),
an n-fold axis (C_n), a center of symmetry (i), or a selection of those elements in combination. In order to be as
definite as possible we shall develop the argument around
a specific example. The ground state of formaldehyde (see
Figure 1) has two symmetry planes, σ_{xz} and σ_{yz}, as well as
a two-fold axis (C_2) along the intersection of the planes.
The existence of these symmetry elements gives rise to a
corresponding number of symmetry operations, which consist
of reflections in one or other of the planes or of rotation

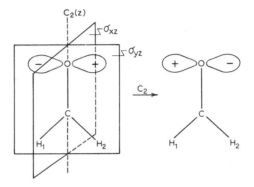

Figure 1   Effect of the operation C_2 upon the n_0 orbital
wave function. C_2 acts on the electron coordinates but not upon the nuclear positions:
its effect is to change the sign of the n_0
function (multiplication by -1).

In this Figure, the z-axis (coincident with
CO axis) is vertical, y is horizontal (in the
plane of the paper) and the x-axis is perpendicular to the plane of the molecule.

around the two-fold axis. These symmetry operations will
be considered to operate on the electron coordinates but not
the nuclear positions: the effect of the operation C_2 on the
n_0 orbital function is then as shown in Figure 1. As the
electron distribution must be unchanged by a symmetry
operation it follows, because the electron distribution is
proportional to the square of the wave function, that the
effect of a symmetry operation is either to leave the wave
function unchanged (multiplication by +1) or change its
sign (multiplication by -1).

Accordingly it is possible to draw up an array of numbers ± 1 encompassing all possible consequences of the symmetry operations applied to the molecule. This is done in Table 1, where the characters ± 1 occur in the body of the table below the symmetry operation to which they refer. It should be noticed that the symmetry operations are not independent of one another: for instance, σ_{xz} and σ_{yz} performed consecutively are equivalent to the single operation C_2. Symbolically, we write

$$\sigma_{xz} \times \sigma_{yz} = C_2 \qquad (6)$$

and the multiplication of the characters must conform. Considerations of this sort limit the number of mutually compatible combinations to those given in the four rows of the table. The table includes an operation E, the <u>identity operation</u>, which consists in leaving the electron coordinates unmoved and whose character is therefore always $+1$. This seemingly trivial operation, introduced in order to satisfy the mathematical concept of a <u>group</u>, is essential in more abstract applications of the theory.

TABLE 1

Character Table for the Group C_{2v}

	E	$C_2(z)$	σ_{xz}	σ_{yz}		
A_1	$+1$	$+1$	$+1$	$+1$	z	
A_2	$+1$	$+1$	-1	-1		R_z
B_1	$+1$	-1	$+1$	-1	x	R_y
B_2	$+1$	-1	-1	$+1$	y	R_x

Each row of characters in the body of Table 1 comprises a symmetry species designated by the shorthand symbol A_1, A_2, ... in the left-hand column. The species A_1, in which the wave function is symmetric (multiplication by $+1$) to every operation, is known as the totally symmetric species: all other species have the property of antisym-

69

metry (multiplication by -1) with respect to two operations and are described as non-totally symmetric. Since the n_0 orbital function of CH_2O changes sign under the operations C_2 and σ_{xz} and is unchanged by E and σ_{yz} its symmetry species is b_2. (We shall use lower case letters for orbital symmetries, reserving capitals for the symmetry species of electronic states.)

Multiplication of the characters of two symmetry species column by column gives a set of characters representing one or other of the four symmetry species of the group. Thus, multiplying the characters of the species B_1 and B_2, one obtains: E, $1 \times 1 = 1$; C_2, $-1 \times -1 = 1$; σ_{xz}, $1 \times -1 = -1$; and σ_{yz}, $-1 \times 1 = -1$. These are the characters of the species A_2, and the result can be summarized by saying that the <u>direct product</u> of the species B_1 and B_2 is the species A_2. Symbolically,

$$B_1 \times B_2 = A_2 \tag{7}$$

Continued multiplication in pairs leads to a table of direct products which is symmetrical about its diagonal so that only half the table need be shown.

Direct Products for the Group C_{2v}

	A_1	A_2	B_1	B_2
A_1	A_1	A_2	B_1	B_2
A_2		A_1	B_2	B_1
B_1			A_1	A_2
B_2				A_1

The contents of the table can in fact be reduced to four simple rules, namely,

$$A \times A = B \times B = A; \quad A \times B = B$$
$$1 \times 1 = 2 \times 2 = 1; \quad 1 \times 2 = 2$$

70

When one electron occupies the n_0 orbital of formaldehyde, its symmetry is b_2: when two electrons are present the symmetry is $b_2 \times b_2 = A_1$. <u>A filled orbital is totally symmetrical</u>, thus we need consider only the singly-occupied orbitals in determining the species of the complete wave function. Thus, the ground electronic state of CH_2O is a totally symmetrical state, A_1, whereas the excited state of the $n \rightarrow \pi^*$ promotion, Equation (2), has orbital symmetry $b_1 \times b_2 = A_2$ and may be either singlet ($^1A_2$) or triplet ($^3A_2$) depending upon the orientation of electron spins in the unfilled orbitals.

To this point we have considered one representative molecule, CH_2O, of one point group, C_{2v}. Details for other point groups are given in references 1 and 2. Structures containing one or more threefold (or higher) axes of symmetry belong to point groups with degenerate symmetry species whose manipulation requires a more detailed theory than that given above.

Intensity of Electronic Transitions

The intensity distribution in an electronic band system is governed by the square of the integral,

$$\int \psi_e^{*\prime} \psi_v^{*\prime} M \psi_e^{\prime\prime} \psi_v^{\prime\prime} d\tau_e d\tau_v \tag{8}$$

in which ψ_e and ψ_v are the electronic and vibrational wave functions and the $'$ and $''$ designate final and initial state of the transition. For electric dipole transitions, the only type now considered, M is the electric dipole moment operator, $\Sigma_i e_i r_i$, where the sum runs over nuclei and elec-

1. F.A. Cotton, "Chemical Applications of Group Theory", Interscience, New York, 1963.

2. D.S. Schonland, "Molecular Symmetry", Van Nostrand, London, 1965.

71

trons. It is customary to simplify the integral (8) by
specifying that the ψ_e refer to the electronic functions
when the nuclei are stationary in their equilibrium positions. In this approximation, the integral factorizes into
two parts,

$$\int \psi_e^{*'} \underset{\sim}{M} \psi_e^{''} d\tau_e \times \int \psi_v^{*'} \psi_v^{''} d\tau_v \qquad (9)$$

of which the first part controls the intensity of the complete band system (the f-value) and the second controls the
distribution of intensity among the individual vibrational
components of the band (the Franck-Condon integral).

The vibrational overlap integral, or Franck-Condon in-
tegral, $\int \psi_v^{'} \psi_v^{''} d\tau_v$ determines the intensity profile in a
band system. In first approximation the functions ψ_v are
products of harmonic oscillator functions $\psi_{v_k}(q_k)$,

$$\psi_v = \psi_1(q_1) \psi_2(q_2) \cdots \psi_k(q_k) \cdots \psi_{3N-6}(q_{3N-6}) \qquad (10)$$

in which the q_k form the set of 3N-6 'normal coordinates'
describing the vibrational motion of the nuclei. Each
normal coordinate is a linear combination of coordinates
for bond extension or bond angle deformation, with coefficients determined by the force constants of the molecule.
Because a molecule has different force constants for each
state the normal coordinates are of different composition;
therefore the Franck-Condon integral is a product of 3N-6
factors, each of the form,

$$\int \psi_k(q_k^{'}) \psi_k(q_k^{''}) dq_k. \qquad (11)$$

the numerical value of such factors being determined by the
geometrical structure and force fields of the excited (')
and ground ('') states of the transition. Sufficient information to evaluate the complete integral is available in
only a few cases. Instead, recourse is often made to ex-
pressly qualitative arguments which recognize that the
vibrations which distort the geometry of one state towards
that of the other will be active in forming progressions in
the spectrum. For a non-totally symmetric transition, the

integrals $\int \psi_k(q_k') \, \psi_k(q_k'') \, dq_k$ are zero or nearly zero, unless $v_k' = v_k''$ these vibrations cannot form progressions. In cases where a change of shape occurs in the transition the classification as totally or non-totally symmetric should be referred to the point group of the state having fewer elements of symmetry.

The statement that an integral is or is not identically zero is called a <u>selection rule</u>. If the integral appearing on the left of the expression (9), i.e. the electronic transition moment, is zero the intensity of the electronic transition must be zero also. Because the value of this integral cannot depend on whether a symmetry operation has been carried out, the integral must be zero unless the integrand, $\psi_e^{*'} \tilde{M}_e \psi_e''$, is totally symmetrical; otherwise one or more symmetry operations exist which will cause the integral to change sign, which is permissible only if its value is identically zero. The symmetry species of the integrand is simply the direct product of the species of its three factors. Now, an electric dipole moment, $\sum_i e_i r_i$, is a vector quantity which must have at least one non-vanishing component along three principal axes xyz. In the C_{2v} group (see Figure 1), the component M_x parallel to x is left unchanged by E and σ_{yz} but reverses direction and therefore sign under the operations C_2 and σ_{xz}: its symmetry species is B_1. The species of M_y (B_2) and M_z (A_1) are determined likewise. Evidently M_x, M_y and M_z have the same symmetry properties as the coordinate axes x, y and z, given in the right-hand column of Table 1. Consider first the allowedness of a transition from the A_1 ground state of CH_2O, having the electron configuration given in Equation (1), to the excited singlet state of the $\pi \rightarrow \pi^*$ transition,

$$\cdots (\pi_{CO}, \, b_1)^1 (n_0, \, b_2)^2 (\pi^*_{CO}, \, b_1)^1: \, {}^{1,3}A_1 \qquad (12)$$

This transition is allowed with a moment change parallel to the CO or z-axis (since the species of ψ_e', M_z and ψ_e'' are then all A_1, and $A_1 \times A_1 \times A_1 = A_1$), but not for moment changes parallel to the x- or y-axes. The direction of the dipole moment change can be determined experimentally,

73

either by analysis of rotational structure in the vapor phase or by observing the directional properties of absorption by molecules held in some definite orientation, as in a single crystal. In fact, the determination of the dipole moment orientation allows one to decide the orbital symmetry of the excited electronic state and so is an essential part of the experimental analysis of an electronic band.

Take as a second example the singlet $n \rightarrow \pi^*$ transition of CH_2O where ψ_e' and ψ_e'' have the species A_2 and A_1, respectively. In order that the transition be allowed, we require a dipole moment operator of species A_2, since $A_2 \times A_2 \times A_1 = A_1$, but no component of the electric dipole moment exists having A_2 symmetry. Accordingly, the singlet $n \rightarrow \pi^*$ transition is <u>electronically forbidden</u>. However, one knows from observation that the transition appears in the absorption spectrum of CH_2O, though with small intensity ($f = 2 \times 10^{-4}$), and it is necessary to seek an explanation.

To the present, the electronic selection rule has been based upon a fixed-nucleus approximation, expression (9), in which the electronic wave function is evaluated for the case when the nuclei are clamped in their equilibrium positions. The approximation is adequate unless the magnitude of the integral $\int \psi_e^* M \psi_e' d\tau$ is zero, or nearly zero, when the errors involved in neglecting the vibrational motion become significant: a transition forbidden for the most symmetrical structure may become weakly allowed when distortion by non-totally symmetrical vibrations is taken into account. Since the allowedness enters only when details of the interaction of <u>vibrational</u> and elect<u>ronic</u> motion are considered, it is said to be caused by <u>vibronic coupling</u>. In electronically-forbidden transitions which owe their appearance to vibronic coupling the 0-0 band, or pure electronic jump, is missing from the spectrum; the observed vibrational transitions are those of the normal Franck-Condon distribution, plus one quantum of a non-totally symmetric vibration. The singlet $n - \pi^*$ band of CH_2O is one example of this behavior, the 250 nm band of benzene forms another. Transitions allowed by vibronic coupling have moderately small f-values, typically $10^{-2} - 10^{-5}$.

It should also be realized that the electronic selection rule depends upon the symmetry of the electronic wave function which, in certain circumstances, may be higher than the symmetry of the nuclear framework. The isotopically-substituted molecule CHDO, for instance, obviously has less symmetry than CH_2O - its point group is C_s rather than C_{2v} - and the distinction is important when vibrational motion (which is strongly dependent on the mass distribution) is considered. But the isotope has a minimal effect on the electronic functions and thus the electronic selection rules for CHDO are unchanged from those for CH_2O. The principle is sometimes extended to the replacement of one substituent group by another, for instance $H \rightarrow CH_3 \rightarrow F$, though the requirement that the electronic wave functions be essentially unchanged by the substitution is always compromised to some extent by a radical change of this type.

Spin-Orbit Coupling

An electron, because of its spin, casts a magnetic field and so responds to external forces like a magnetic dipole or small bar magnet. In its orbital motion the electron continuously cuts the lines of force of the nucleus' electric field; in consequence, each interacts with the other in a manner quite separate from the ordinary electrostatic attraction of oppositely-charged particles. This form of interaction is known as spin-orbit coupling.

In a triplet state, the symmetry of the complete wave function, $\psi_e\psi_{spin}$, is given by the direct product of the symmetry species of the orbital function and that of the spin function. Triplet spin functions have the same symmetry properties as a rotation of the whole molecule; thus in the group C_{2v} they have the species B_2, B_1 and A_2, corresponding to the three rotations R_x, R_y and R_z (right-hand column of Table 1). The singlet spin function however is totally symmetrical, so that for singlet states $\psi_e\psi_{spin}$ has the same symmetry as ψ_e itself. Spin-orbit coupling weakly mixes singlet and triplet states for which the sym-

metries of $\psi_e \psi_{spin}$ is the same for both. Consider, for
example, that the lowest triplet state T_1 may mix in this
fashion with some state S_i of the singlet manifold to which
transitions from the S_o ground state are allowed in absorption. The very small amount of T_1 present in the wave
function for the S_i state has essentially no effect upon
the $S_o \rightarrow S_i$ transition, but the small admixture of S_i into
T_i provides a mechanism by which the otherwise total prohibition on change of electron spin can be overcome, so
that the $S_o \rightarrow T_1$ transition occurs weakly in the spectrum.
Since spin-orbit coupling is small in organic molecules,
singlet-triplet transitions have very low intensities. For
molecules built from atoms from the first row of the periodic table the upper limit to the f-value of singlet-
triplet transitions is about 10^{-7}. Atoms in higher rows of
the table (e.g. Br, I) introduce stronger spin-orbit coupling and some intensification of S-T absorption may then be
observed.

Take for example the $n \rightarrow \pi^*$ triplet transition of
formaldehyde. The triplet state has orbital A_2 symmetry
and the spin functions have species B_2 (R_x), B_1 (R_y) and
A_2 (R_z); thus the symmetry of the total wave function is
$A_2 \times B_2 = B_1$, or $A_2 \times B_1 = B_2$, or $A_2 \times A_2 = A_1$. The triplet state therefore mixes by spin-orbit coupling with sing-
let states of B_1, B_2 or A_1 symmetry, the relative importance being a matter for observation or calculation. It
turns out that the most significant mixing is with the A_1
state of the $\pi \rightarrow \pi^*$ excitation, i.e., the state whose electron configuration is given by Equation (12), and that this
mixing is relatively strong so that the $S_o \rightarrow T_1$ transition
of formaldehyde, and carbonyl compounds generally, is one of
the most intense singlet-triplet bands observed for organic
structures.

Applications

In this section we shall review some of the known facts
pertaining to the C=C, C=O and C_6H_6 chromophores. A summary
of this information is presented in the table at the end of
the chapter. Usually this information is much more complete

for the first member of each series than for its derivatives or homologs.

Chemists are accustomed to energy relationships in
terms of kilocalories per mole. In dealing with spectra,
according to the number of significant figures one wishes
to convey, it is often more convenient to use the per molecule units of electron-volts (eV) or wavenumber (cm^{-1}).
The conversion table for the three sets of units is given
below.

TABLE 2

	kcal mole^{-1}	cm^{-1}	eV
1 kcal mole^{-1}	1	349.8	4.336×10^{-2}
1 cm^{-1}	2.859×10^{-3}	1	1.240×10^{-4}
1 eV	23.06	8066	1

Wavelengths are here expressed in nanometers (nm): the
corresponding wavenumber in units of cm^{-1} is 10^7 times the
reciprocal wavelength, $(nm)^{-1} \times 10^7$.

The C=C Double Bond

(a) C_2H_4 Ethylene[3] has three distinct absorptions
in the nearer ultraviolet region. The first absorption, an
exceedingly weak band in the region 270-350 nm, is recognized as a singlet-triplet transition by the fact that its
intensity increases markedly in the presence of O_2. The

---

3. A.J. Merer and R.S. Mulliken, Chem. Revs., 69,
639 (1969).

absorption was discovered using 1.4 m path of liquid ethylene, thus the peak molar absorptivity is not greater than about 10^{-4}: even so, it is probable that its observation depended upon the incomplete removal of oxygen and that the intrinsic intensity is smaller still. Steeply-rising absorption at 265 nm marks the beginning of the second absorption which forms an extremely broad band with its maximum near 162 nm (ε_{max} ~10^4) though the envelope can be followed to below 140 nm: the f-value for this transition is 0.34. The third absorption is somewhat weaker (f~0.04) and falls on the rising shoulder of the second absorption, near 175nm. This system has sharp vibrational structure, quite distinct from the almost continuous envelope of the second absorption (see Figure 2). A fourth system was formerly thought to be present near 200 nm - the so-called 'mystery' bands of ethylene - but it is now known that the evidence for this system is spurious.

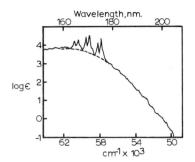

Figure 2   Envelope of the overlapping singlet $S_0 \rightarrow S_1$ ($\pi \rightarrow \pi^*$) and $S_0 \rightarrow S_2$ (Rydberg) transitions of C_2H_4 vapor [Wilkinson and Mulliken, reference 4].

4.   P.G. Wilkinson and R.S. Mulliken, J. Chem. Phys., 23, 1895 (1955).

In order to interpret this pattern of bands it is
necessary to say something about the molecular orbitals of
ethylene. The highest-energy M.O. of <u>planar</u> ethylene occupied in the electronic ground state is the bonding π_{CC} orbital, whilst the lowest vacant orbitals are the antibonding
π^*_{CC} and CH* orbitals. Regarding all except the 1s electrons of carbon like those of a single atom, we find that
the CH* orbital behaves like a 3d atomic orbital of the
"semi-united" atom Si, thus it will be appropriate to regard CH* as an extra-valence, or <u>Rydberg</u>, orbital. These
three orbitals, with symmetry labels appropriate to D_{2h}
(planar) ethylene, are shown in Figure 3. The $\pi \rightarrow \pi^*$ one-
electron promotion can be expected to give rise to the triplet and singlet states T_1 and S_1, while the higher-energy
$\pi \rightarrow CH^*$ promotion should yield triplet and singlet

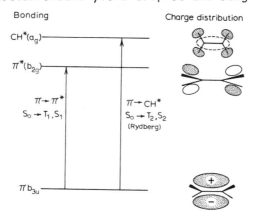

Figure 3   Molecular orbitals for ethylene (adapted from
             Merer and Mulliken[5]). The orbital description
             appears on the left side together with the
             symmetry classification for planar ethylene
             (D_{2h} group): a qualitative impression of the
             charge distribution in each orbital appears
             on the right side. Each vertical line repre-
             sents a one-electron promotion, therefore to
             two electronic transitions since the final
             (excited) state may be either a singlet or a
             triplet state.

5.   A.J. Merer and R.S. Milliken, <u>Chem. Revs.</u>, 69,
       639 (1969).

Rydberg states. If it is accepted that the triplet Rydberg transition is not observed, it is not difficult to imagine that the three remaining promotions give rise to the very strong and broad 162 nm absorption (singlet $\pi \to \pi^*$), the medium intense 175 nm absorption (singlet $\pi \to CH^*$), and the extremely weak 300 nm absorption (triplet $\pi \to \pi^*$). It turns out that this simple scheme is correct, but that the details are quite subtle and complex.

Consider first what happens to the S_0 and S_1 states of planar ethylene when the molecule is twisted. Although the π and π^* M.O.'s are respectively strongly bonding and strongly antibonding in planar ethylene, they both become non-bonding for 90° twisted ethylene: thus one part of the carbon-carbon double bond, the π bond, vanishes. (However, to some degree it is replaced by hyperconjugation, so that 90° twisted ethylene is not quite as unstable as one might have otherwise thought.) The twisted configuration also shows a close correlation of its orbitals with those of the iso-electronic molecule O_2, which has a triplet ground state, and when this relationship is explored quantitatively, it turns out that the lowest state of twisted ethylene must also be a triplet state. Because the energy of the lowest <u>singlet</u> state of twisted ethylene can be identified with the activation energy (65 kcal) for thermal cis-trans isomerization of $CHD \cdot CHD$, we may conclude that the T_1 state of ethylene is 90° twisted and that its energy is not more than 65 kcal (2.8 eV) higher than that of the planar S_0 state. For the same reason, namely, relief of antibonding, the first excited singlet state S_1 is also expected to be 90° twisted in its equilibrium configuration.

The peak absorption intensity in the $S_0 \to T_1$ and $S_0 \to S_1$ systems does not, of course, correspond with the transition between vibrationless levels: instead, it marks the position of the vertical transitions which, according to the Franck-Condon principle, go to upper state levels lying at the top of the barrier to internal rotation, where the wave function of the twisting motion has the greatest amplitude at the planar configuration. Strictly, the vertical transition involves some excitation of the C-C stretching vibration also, since the bond distances in the S_1 and T_1

80

states differ from that of the S_o ground state. At present only a qualitative analysis of the Franck-Condon envelope of the $S_o \rightarrow T_1$ and $S_o \rightarrow S_1$ transitions has been carried out, thus the structures of the T_1 and S_1 states are not precisely tied down. Nevertheless the equilibrium C-C distances in T_1 and S_1 are believed to be about 1.58 Å and 1.44Å respectively, and the barrier to internal rotation in both states is estimated to be 46 ± 9 kcal (2 ± 0.4 eV), significantly less than the 65 kcal barrier opposing isomerization in the ground state. It should be stressed however that the barrier in S_1 and T_1 states represents the excess energy of the unstable, planar configuration over that of the stable 90°-twisted form, the converse of the situation that applies in the ground state. The S_1 state lies about 4.7 eV above S_o, corresponding to an 0-0 transition at 265 nm.

In the third state, the singlet Rydberg state S_2, the excited electron is promoted to an orbital of large radius, so that the state represents approximately an intermediate stage between the ground state of ordinary ethylene and the ground state of the $C_2H_4^+$ ion. One might at first expect that $C_2H_4^+$, with one π electron removed, will have a planar structure with perhaps half the barrier opposing internal rotation as ethylene itself. But this is not the case: all theoretical treatments are unanimous that $C_2H_4^+$ must have a partially-twisted structure in which the preference of the single π electron for a planar configuration is offset by the gain in hyperconjugation resulting from a twisted configuration. (The hyperconjugation brought into play here is considerably more potent than the mild form realized in, for example, propylene or toluene.) The Rydberg system has considerably more structure than the two systems discussed earlier and the experimental conclusions are correspondingly more definite. Thus, the Rydberg state of ethylene has a C-C distance of 1.41 ± 0.01 Å, close to that in the ordinary ground state of benzene (1.397 Å) where the CC bond is also a 3-electron bond. At equilibrium, the CH_2 planes are inclined at angles of $25 \pm 1°$ to the mean plane so that an energy barrier must be surmounted to bring the molecule into either the planar or the 90° twisted configuration.

The lesser barrier, 290 cm^{-1} or 0.83 kcal, is that opposing twisting into the planar configuration: the barrier to the $90°$ twisted configuration is much greater, 3340 cm^{-1} (9.5 kcal). Both are significantly smaller than the barriers of 46 ± 9 kcal in the S_1 or T_1 states, or of 65 kcal in the S_0 state, a visible effect of the competition between π bonding and hyperconjugation in the Rydberg state. Because the geometry change is comparatively mild, the 0-0 band - or pure electronic jump - is prominent in the $S_0 \rightarrow S_2$ transition and can easily be identified from the envelope shown in Figure 2 as the sharp, strong peak at 57340 cm^{-1} (164 kcal/mole).

(b) <u>Substituted Ethylenes</u>. The absorption of alkyl-ethylenes, as expected, has a strong likeness to that of ethylene itself. Faint S_0 - T_1 bands are observed in presence of added oxygen, though the experimental difficulties are such that this has **not been** accomplished in many examples. The S_0 - S_1 ($\pi \rightarrow \pi^*$) band, which retains its characteristic, broad envelope and moderately high intensity ($\varepsilon_{max} \sim 10^4$), is shifted progressively to the red by alkyl substitution so that the peak wavelength increases from 162 nm in ethylene itself to about 188 nm for tetramethyl-ethylene. The unseen 0-0 band must be similarly affected, but there is no firm spectroscopic information as to the magnitude of the shift.

The Rydberg transition responds more to alkyl substitution than either of the π - π^* transitions. This transition shifts rapidly toward longer wavelengths and, in fully alkylated ethylenes, is marked by a distinct medium-intense band ($\varepsilon = 500-1000$), separated from the flank of the S_0 - S_1 envelope, at about 235 nm. The 0-0 band of this transition is usually quite prominent in vapor spectra. Most substituted ethylenes also show a definite shoulder ($\varepsilon_{max} \sim 0.1$) displaced 30-35 nm to the red of the Rydberg band which probably represents the transition to the triplet Rydberg state, that is, to the fourth state predicted from the simple orbital picture in Figure 3. This state is unobserved in ethylene itself, possibly because the absorption lies buried under the S_0 - S_1 envelope.

The Carbonyl Group

   (a) CH_2O. Formaldehyde[6] has the same number of electrons as ethylene from which it can be formed, imagina-
tively, by amalgamating the two protons and the carbon
nucleus. In this process four CH bonding electrons are
converted into two nonbonding pairs, one of which lies deep
in the orbital manifold and plays no part in the optical
absorption. The other pair occupies a b_2 orbital intermediate in energy between the bonding π and antibonding π^*
orbitals. Otherwise, the organization of filled and unfilled orbitals is similar to that previously shown for
ethylene in Figure 3; the optical transitions are those of
ethylene plus the singlet and triplet n → π^* transitions
characteristic of all carbonyl compounds.

   The first electronic band system of formaldehyde
commences near 395 nm and marks the triplet n → π^* excitation. Although this transition must be classed as weak
$(\varepsilon \sim 0.1 - 0.01)$ it is nevertheless strong in comparison with
many singlet-triplet transitions and is quite easy to
observe experimentally. Its intensity is derived by spin-
orbit mixing of the triplet state with the singlet upper
state of the π → π^* electron promotion, that is to say,
with the state analogous to the S_1 state of ethylene. Although this mixing is well-established, theoretically and
experimentally, very little else is known about the π, π^*
excited singlet state of formaldehyde except that it is
believed to be considerably higher in energy than the
corresponding state of ethylene.

   The second absorption of formaldehyde covers the region 350-250 nm and represents the singlet n → π^* electron
promotion. It is therefore considerably more intense than

---

6.   J.K. Sidman, Chem. Revs., 58, 689 (1958).

the first absorption ($\epsilon_{max} \sim 10$; $f \sim 2 \times 10^{-4}$) though still a
rather weak transition in absolute terms. The reason for
its low intensity is that the pure electronic jump is forbidden as an electric dipole transition, that is to say, it
cannot occur through the interaction of the undistorted
molecule with the electric field of the radiation. It is,
in fact, allowed to occur through interaction with the
magnetic field of the radiation, but this is of relatively
minor importance and the band system owes most of its intensity to a mechanism of vibronic coupling (see p. 74, this
chapter), whereby the transition "borrows" intensity from
strong transitions lying at higher energy, in this case,
transitions in the far ultraviolet. The mechanism, however,
is relatively inefficient and the intensity remains low.

Both singlet and triplet states of the n - π transition have similar structures. The CO bond distance is long
(1.31 Å in the triplet, 1.32 Å in the singlet state), as
might be expected from the fact that the π part of the bond
now encompasses two bonding and one antibonding electron.
As usually happens, the antibonding effect is stronger than
that of bonding (see Chapter 2), thus the CO distance is
somewhat longer than that associated with a 3-electron bond
(1.27 Å in the ion $HCOO^-$). The most characteristic feature
of the states is that they are also non-planar, the final
structure being a compromise between relief of antibonding
- accomplished by bending the molecular plane - and the
energetic preference of the partial π bond for a planar configuration. In the singlet state the angle between the CO
bond and the CH_2 plane is about 20°, in the triplet state
about 35°; the barriers opposing planarity are small, respectively 1.0 and 2.2 kcal[7,8]. The structure can be vis-
ualized as derived from methyl alcohol by pulling off two
hydrogen atoms, then allowing the pyramid to relax about

---

7.   W.T. Raynes, J. Chem. Phys. 41, 2755 (1966);
       V.T. Jones and J.B. Coon, J. Mol. Spectrosc. 31,
       137 (1969).

8.   V.A. Job, V. Sethuraman and K.K. Innes, J. Mol.
       Spectrosc. 30, 365 (1969).

half-way to a planar configuration. The triplet state of
the $\pi \to \pi^*$ excitation is not known spectroscopically.

   (b)   Unsaturated Carbonyl Compounds[9]. The absorption
can be classified in the manner used for saturated carbonyl
compounds, as singlet-singlet or singlet-triplet $\pi \to \pi^*$ or
$n \to \pi^*$. Conjugation has only a mild effect on the $n \to \pi^*$
singlet transitions, which are displaced slightly to longer
wavelengths so that the 0-0 band of trans-propenal occurs at
386 nm compared to 355 nm for CH_2O. The $\pi \to \pi^*$ transition,
however, moves sharply to longer wavelengths, into the conventional near-ultraviolet region, and its position can be
predicted quite accurately using Woodward's empirical[10]
rules. Since this transition is thought to involve a considerable reorganization of π electron density, it is sometimes described as a 'charge-transfer' transition.

   Unfortunately, not much experimental information is
available. No information exists on the singlet state
structure resulting from the $\pi \to \pi^*$ excitation. The
singlet-triplet $n \to \pi^*$ excitation, though weak, is quite
easily observed, but no detailed analysis of the bands has
been achieved and the only conclusion is that, in a general
way, the physical structure of the excited state probably
resembles that of the corresponding singlet state. The
latter, though known in outline only, can be understood by
reference to the schematic diagram of the π^* orbital in
Figure 4. This orbital is bonding between the central pair
of atoms in the skeleton and antibonding between outer
pairs. It is consistent with this picture that, compared
with the ground state, the barrier opposing internal rotation about the central C-C bond in trans-propenal is substantially higher in the $n - \pi^*$ excited state, whilst the

---

9.   J.C.D. Brand and D.G. Williamson, Disc. Faraday Soc.
       35, 184 (1963); J.M. Hollas, Spectrochim. Acta 19,
       1425 (1963).

10.   For illustrations of the use of Woodward's rules,
       see e.g. "Technique of Organic Chemistry",
       A. Weissberger Ed., Ch. II, Vol. XI, Interscience,
       New York, 1963.

barrier opposing rotation about the ethylenic C-C bond is substantially lowered. Unlike the equivalent state of CH_2O the structure is planar, corresponding to a roughly equal mixture of the valence-bond structures,

$$\begin{array}{cc}
\begin{array}{c}
\diagup \dot{C} - C \diagdown \\
\diagdown \\
\quad\quad C - \bar{O}\cdot \\
\diagup
\end{array}
& \text{and} &
\begin{array}{c}
\diagup C = C \diagdown \\
\diagdown \\
\quad\quad \dot{C} - \bar{O}\cdot \\
\diagup
\end{array}
\end{array}$$

In the ground state, <u>cis</u>-propenal is less stable than the <u>trans</u>-conformation by about 700 cm^{-1} or 2.0 kcal/mole[11]. The relative order is reversed in the singlet and triplet states of the n → π^* excitation where the <u>cis</u>-conformation is more stable than the <u>trans</u>- by 530 cm^{-1} and 420 cm^{-1}, respectively.

Figure 4   Charge distribution in the first π^* orbital of propenal. This orbital is bonding between the central carbon atoms but antibonding with respect to the CO and terminal CC pairs. A higher π^* orbital (not shown) exists which is antibonding with respect to all neighbors in the skeletal chain.

---

11.   E.J. Bair, W. Goetz and D.A. Ramsay, <u>Can. J. Phys.</u>, 49, 2710 (1971).

Aromatic Compounds

(a)   Benzene, Naphthalene.   Benzene has been more intensively studied than any other organic compound.   The first four absorption regions known in the ultraviolet have the characteristics given in the following table.

TABLE 3

Benzene

λ, nm	340	260	205	180
E_o, kcal	84	109	139	159
f	$< 10^{-11}$	0.002	0.12	1.2
Assign-ment	$S_o \rightarrow T_1$	$S_o \rightarrow S_1$	$S_o \rightarrow S_2$	$S_o \rightarrow S_3$
	$^1A_{1g} \rightarrow {}^3B_{1u}$	$^1A_{1g} \rightarrow {}^1B_{2u}$	$^1A_{1g} \rightarrow {}^1B_{1u}$	$^1A_{1g} \rightarrow {}^1E_{1u}$

The transitions observed have for many years been inter-preted on the assumption that the excitation involves only the π electrons.   The one-electron π orbitals of benzene are classed a_{2u}, e_{1g}, e_{2u} and b_{2g} in order of increasing energy.   The first excited electronic configuration,

$$(a_{2u})^2 (e_{1g})^3 (e_{2u})^1 \; : \; {}^{1,3}B_{2u}, \; {}^{1,3}B_{1u}, \; {}^{1,3}E_{1u} \qquad (13)$$

then gives rise to singlet and triplet states of symmetry species B_{2u}, B_{1u} and E_{1u}.   It is agreed theoretically that the singlet states occur in the order given, that is, $^1B_{2u}$ should be identified with S_1, $^1B_{1u}$ with S_2, and $^1E_{1u}$ with S_3.   In the triplet manifold, all calculations are unani-mous that the lowest state T_1 is $^3B_{1u}$.   Of the states named only $^1E_{1u}$ (S_3) can combine with the ground state in an electronically-allowed transition.

87

The intense 180 nm band corresponds with the allowed $^1A_{1g} \to {}^1E_{1u}$ transition. The two singlet-singlet transitions at lower energy occur through vibronic coupling, their intensity being 'borrowed' from the 180 nm band. Not much is known concerning the physical geometry of the S_2 and S_3 states but the S_1 state has the same planar, regular hexagonal structure as the ground state, though the C-C distance (1.435 Å) is significantly larger - it is 1.397 Å in the ground state - corresponding to an increase in antibonding.

The 340 nm bands are not observed in absorption except in presence of oxygen, but phosphorescence from this state can be measured in rigid media at low temperature. Although the ESR spectrum of triplet benzene in the solid phase indicates that distortion occurs from the regular hexagonal geometry, it is not yet certain whether this is a property of the isolated molecule or whether the distortion is enforced by the environment. Part of the explanation for the very low intensity of the $S_o \to T_1$ absorption is that spin-orbit coupling mixes the $^3B_{1u}$ state with $^1B_{2u}$, but not with $^1B_{1u}$ or $^1E_{1u}$, so that the $S_o \to T_1$ transition draws its intensity from S_o - S_1 which is itself electronically forbidden and therefore weak.

Naphthalene and higher aromatic hydrocarbons closely resemble benzene in that the electronically-allowed transition is preceded in the spectrum by weaker, forbidden transitions. Data for naphthalene, summarized in the table below show an almost one-to-one correlation with benzene apart from the wavelength shift.

TABLE 4

Naphthalene

λ, nm	470	320	278	220
f	$\sim 10^{-9}$	0.002	0.18	1.7
Assignment	$S_o \to T_1$	$S_o \to S_1$	$S_o \to S_2$	$S_o \to S_3$

The aza-derivatives of benzene and naphthalene (pyridine, pyrazine, ... , quinoline, etc.) show both $\pi \to \pi^*$ type absorption, generally quite closely correlated with that of the parent hydrocarbon, plus $n \to \pi^*$ absorption at longer wavelengths. A comprehensive review of these spectra has been published recently[12].

(b) Monosubstituted Derivatives. Substituents perturb the π-electron spectrum of benzene to a degree which depends on the nature of substituent: methyl has only a minor effect, unsaturated groups (e.g. NO_2) and substituents carrying lone-pair electrons (OH, NH_2) produce much larger changes. The information available refers mainly to those states which correlate with the T_1 ($^3B_{1u}$) and S_1 ($^1B_{2u}$) states of benzene, and only these states are considered here.

In the S_1 states the interaction of the substituent with the π-electrons of the ring is much stronger than in the ground state. Even with a 'strong' substituent (OH, NH_2), the phenyl group remains essentially a regular hexagon in the electronic ground state of these molecules, whereas in the S_1 state the ring is definitely distorted from hexagonal symmetry. In this sense, all ground state phenyl-substituent interactions should be classed as mild relative to the interaction that occurs in the S_1 state. The reorganized charge distribution can be seen also from the change in acidity of hydroxy- and amino-derivatives of aromatic hydrocarbons. The S_0 state of a phenol is mildly acidic, the S_1 state has similar properties to formic acid. Data for some β-naphthyl derivatives are given in the table at the end of this section[13]. Similar results available for the T_1 states show the acidic dissociation constants in

---

12.   K.K. Innes, J.P. Byrne and I.G. Ross, J. Mol. Spectrosc., 22, 125 (1967).

13.   G. Jackson and G. Porter, Proc. Roy. Soc., A, 260, 13 (1961); T.S. Godfrey, G. Porter and P. Suppan, Disc. Faraday Soc., 39, 194 (1965).

this state are much closer to ground state constants than to those of the S_1 state. Evidently the electron distribution in the triplet state is more like that in S_0 than in S_1.

TABLE 5

Compound	$pK_a(S_0)$	$pK_a(S_1)$	$pK_a(T_1)$
2-Naphthol	9.5	3.1	8.1
2-Naphthoic acid	4.2	>10	4.0
2-Naphthylammonium cation	4.1	-2	3.3

Oxygen

The lowest-energy electronic configuration of molecular oxygen is

$$\ldots\ldots (\pi_u)^4(\pi_g)^2 \quad : \quad {}^3\Sigma_g^-, \ {}^1\Delta_g, \ {}^1\Sigma_g^+ \qquad (14)$$

The π bond therefore comprises six electrons, four of which occupy the bonding π_u orbital while the remaining two are present in the half-filled antibonding orbital π_g. To the extent that the antibonding electrons roughly cancel the bonding power of the two bonding π electrons the net bonding is contributed by two σ electrons and two π electrons, corresponding approximately to a formal double bond.

Because the orbital and spin angular momenta of the electrons in the half-filled π_g orbital can be aligned in different ways the configuration (14) corresponds to three distinct electronic states which, in order of increasing energy are ${}^3\Sigma_g^-$ (ground), ${}^1\Delta_g$ (0.98 eV) and ${}^1\Sigma_g^+$ (1.64 eV). The electron distribution of course is essentially the same for all three states so that the internuclear distance shows only a mild variation, from 1.207 Å for ${}^3\Sigma_g^-$ to 1.227 Å for ${}^1\Sigma_g^+$. Transitions from the ground state to the

90

TABLE 6

Electronic Transitions and States of Some Organic Molecules

Molecule	State	Point Group	E_o kcal	Structure	Transition λ(nm)	f-Value	Vibrational structure (cm^{-1})	Comment
C_2H_4 [1]	$S_o(^1A_{1g})$	D_{2h}	0	rCC=1.335 ν_{CC}" 1623	$S_o \rightarrow T_1$ (260-350)			Observed in presence of O_2. T_1 state is 90° twisted. Barrier to planarity 37-55 kcal.
	$T_1(^3A_2)$	D_{2d}	<65	rCC~1.6				
	$S_1(^1B_2)$	D_{2d}	~108	rCC~1.44	$S_o \rightarrow S_1$ (140-215)	0.34	Intervals of 700-950 cm^{-1} probably involve superposed progressions in CC-stretching and twisting vibrations.	S_1 state is 90° twisted. Barrier to planarity 37-55 kcal.
	$S_2(^1B_3)$	D_2	164	rCC~1.41	$S_o \rightarrow S_2$ (155-175)	0.04	ν_{CC}' 1380 Irregular progression in twisting motion.	S_2 state 25° twisted. Barriers of 0.8 and 9.5 kcal.
CH_2O [2]	$S_o(^1A_1)$	C_2	0	rCO=1.208 ν_{CO}" 1745	$S_o \rightarrow T_1$ [3] (350-420)	10^{-7}	ν_{CO}' 1280 Irregular progression in inversion mode.	T_1 state is pyramidal with out-of-plane angle ~35°. Barrier to planarity 2.2 kcal. Spin-orbit coupling with higher $^1A_1$ ($\pi \rightarrow \pi*$) state.
	$T_1(^3A_2)$	C_s	72	rCO=1.31	$S_o \rightarrow S_1$ [4]	2×10^{-4}	ν_{CO}' 1180 Irregular progression in inversion mode.	S_1 state is pyramidal with out-of-plane angle ~20°. Barrier to planarity 1.0 kcal. 0-0 band forbidden for electric-dipole selection rules.
	$S_1(^1A_2)$	C_s	80.5	rCO=1.321				

1. A.J. Merer and R.S. Mulliken, Chem. Revs. 69, 639 (1969).

2. J.K. Sidman, Chem. Revs. 58, 689 (1958).

3. W.T. Raynes, J. Chem. Phys. 41, 2755 (1966); V.T. Jones and J.B. Coon, J. Mol. Spectrosc. 31, 137 (1969).

4. V.A. Job, V. Sethuraman and K.K. Innes, J. Mol. Spectrosc. 30, 365 (1969).

Molecule	State	Point Group	E_0 kcal	Structure	Transition λ(nm)	f-Value	Vibrational structure (cm^{-1})	Comment
$CH_2{:}CH{\cdot}CHO$ (trans)[1]	$S_0(^1A')$	C_s	0					T_1 state (upper state of n → π* promotion) is planar.
	$T_1(^3A'')$	C_s	68		$S_0 - T_1$ (390–412)			S_1 state is planar, with higher barrier to rotation about C·C than S_0.
	$S_1(^1A'')$	C_s	74		$S_0 \rightarrow S_1$ (290–390)		ν_{CO}' 1266 ν_{CC}' 1410	
	S_2	?	~140		$S_0 \rightarrow S_2$ (180–230)			π → π* excitation, with probable structural changes.
$HC{:}C{\cdot}CHO$[2]	$S_0(^1A')$	C_s	0		$S_0 \rightarrow S_1$ (280–390)	$\sim10^{-4}$		S_1 state planar or nearly planar (max. out-of-plane angle, 4°).
	$T_1(^3A'')$	C_s	69					
	$S_1(^1A_1'')$	C_s	75					
	Higher singlet states near 115 and 140 kcal							
C_6H_6	$S_0(^1A_{1g})$	D_{6h}	0	rCC=1.397 ν_1'' 995 ν_6'' 608 (ring) ν_8'' 1596	$S_0 - T_1$ (310–340)	$<10^{-11}$	ν_1', ~900(ring)	Observed in presence of O_2. T_1 state planar, possibly not regular, hexagon. Spin-orbit coupling allowed with S_1 state. 0–0 band forbidden
	$T_1(^3B_{1u})$	D_{6h}	84					
	$S_1(^1B_{2u})$	D_{6h}	109	rCC=1.434	$S_0 \rightarrow S_1$ (220–265)	0.002	ν_1' 923 ν_6' 522	S_1 state is a planar regular hexagon. 0–0 band forbidden. ν_6 (ring deformation) active in vibronic coupling.

1. J.C.D. Brand and D.G. Williamson, Discussion Faraday Soc. 35, 184 (1963).
    J.M. Hollas, Spectrochim. Acta 19, 1425 (1963).

2. J.C.D. Brand, J.H. Callomon and J.K.G. Watson, Discussions Faraday Soc. 35, 175 (1963).

3. J.H. Callomon, T.M. Dunn and I.M. Mills, Phil. Trans. Roy. Soc. London, Ser. A, 259 (1966).

Molecule	State	Point Group	kcal	Structure	Transition λ(nm)	f-Value	Vibrational structure (cm^{-1})	Comment
	$S_2(^1B_{1u})$	D_{6h}	139		$S_o \to S_2^1$ (190–205)	0.12	ν_1' 920±50 (rare gas matrix)	S_2 state is planar hexagonal. 0–0 band forbidden. Vibronic coupling involves mainly ν_8 (CC stretching).
	$S_3(^1E_{1u})$	D_{6h}	159		$S_o \to S_3^1$ (175–190)	1.2	ν_1' 920±50 (rare gas matrix)	S_3 structure is planar hexagonal. Transition is electronically allowed.

1. B. Katz, M. Brith, B. Scharf and J. Jortner, J. Chem. Phys. 52, 88 (1970).

$^1\Sigma_g$ and $^1\Delta_g{}^+$ states are opposed by the general prohibition on change of spin and thus have the characteristically low intensity of a spin-forbidden transition. Since the for-bidden-ness applies equally to the upward transition occurring by absorption of radiation, or the downward transition associated with emission, the radiative lifetimes of the $^1\Sigma_g{}^+$ and $^1\Delta_g$ states are correspondingly long, 7 sec. and 45 min., respectively. These lifetimes represent the behavior under conditions of very low pressure, as in the outer atmosphere. At higher pressures, as in the condensed phase the actual lifetime is controlled by bimolecular processes.

The first strong absorption by molecular O_2, which commences near 200 nm, marks a $\pi \rightarrow \pi^*$ type of excitation in many respects analogous to the singlet $\pi \rightarrow \pi^*$ excitation of C_2H_4.

Oxygen molecules have the useful property of intensi-fying $S_o \rightarrow T_1$ absorption by other molecules. The effect may be due to formation of transient complexes which, for aromatic substates, can be formulated (aromatic)$_m$ - $(O_2)_n$ where m usually equals 1 and n usually equals 1 or 2. Absorption by these complexes can be detected in the ultra-violet spectrum of solutions of oxygen in aromatic hydro-carbons, alongside the normal or intensified absorption of the aromatic group.

CHAPTER 4

TRANSIENTS AND THEIR BEHAVIOR

Types of Transients

Most photochemical reactions are complex in their mechanisms and involve a number of reactive intermediates or transients. Thus, an understanding of the nature, lifetime and reactions of these transients is a prerequisite to the understanding of the overall mechanism.

The first type of transient in any photochemical reaction must be the electronic excited states of the absorbing molecule. As noted in Chapter 1, these excited states can be either of <u>singlet</u> or <u>triplet</u> character, and photochemistry can occur from either. These photoexcited states are usually detected by monitoring the emission from the singlet state (fluorescence) or the triplet state (phosphorescence). They may also be detected by direct absorption, an experiment that is more difficult with singlets than triplets.

Other transients encountered are the intermediate reaction products resulting from the decomposition or chemical quenching of the excited states. Such reactions are called <u>primary photochemical reactions</u>. Several kinds of reaction leading to chemical transients can be discerned; each of these processes will be considered in more detail later.

In this chapter we begin with a discussion of the characteristics of excited states, and then consider in

detail how these excited states react to form chemical
transients.

Lifetimes of Excited States

As noted in Chapter 1 the radiative or natural lifetime τ_0 of an excited state is defined as the reciprocal
of the rate constant for radiative emission from that excited state provided that spontaneous emission is the only
mode of decay. τ_0 is given by

$$\tau_0 = \frac{1}{A_{no}}$$

where A_{no} is the Einstein probability for spontaneous
emission from the excited state $\underline{n}$ to the ground state $\underline{o}$.
A_{no} is directly proportional to the Einstein probability
of induced absorption B_{on}; hence, τ_0 is entirely determined
by the absorption intensity of the transition. If the
absorption band is narrow, then τ_0 may be approximated by[1]

$$\tau_0 = \frac{g_n N}{g_0 8\pi\tilde{\nu}_{no}^2 n^2 2303 c} \bigg/ \int \epsilon d\tilde{\nu} \qquad (1)$$

Here g_n and g_0 are the degeneracies of the excited state
and ground state respectively, N is Avogadro's number,
$\tilde{\nu}_{no}$ (in cm^{-1}) is the center frequency of the transition,
c is the speed of light, n is the refractive index and the
integral is taken over the absorption band corresponding
to this state of interest. However, as is usually the
case, the absorption band is broad and a modified form of
Equation (1), as derived by Strickler and Berg[2], must be

1.   J.G. Calvert and J.N. Pitts, Jr., "Photochemistry"
      John Wiley and Sons, Inc., New York, 1966, p. 174.

2.   S.J. Strickler and R.A. Berg, J. Chem. Phys.,
      37, 814 (1962).

employed

$$\tau_o = \frac{g_n N <\widetilde{\nu}_f^{-3}>_{Av}}{g_o 8\pi n^2 2303 c} \Bigg/ \int \epsilon d\ln \nu \qquad (2)$$

The symbols are defined as for Equation (1) except that

$$<\widetilde{\nu}^{-3}> = \frac{\int \widetilde{\nu}^{-3} (\widetilde{\nu}) d\widetilde{\nu}}{\int (\widetilde{\nu}) d\widetilde{\nu}} \qquad (3)$$

where $I(\widetilde{\nu})$ is the relative intensity (in quanta) of the fluorescence spectrum at the frequency $\widetilde{\nu}$. Equation (2) takes into account the proper averaging of the $\widetilde{\nu}^{-3}$ term over the spectrum. Comparisons of observed lifetimes vs. lifetimes calculated from Equation (2) show good agreement provided that the transition is an allowed one and there is a good mirror image relationship between the absorption and emission spectra.

As a very rough approximation, for absorption bands with a half width of ~ 5000 cm^{-1} and centered at $\sim 30,000$ cm^{-1} in the ultraviolet

$$\tau_o \text{ (sec)} \sim \frac{10^{-4}}{\epsilon_{max}} \qquad (4)$$

For strongly allowed transitions ($\epsilon_{max} \sim 10^4$) τ_o will be ~ 10 nanoseconds or less. On the other hand very weakly allowed transitions (such as singlet-triplet transitions where ϵ_{max} can be 10^{-4} or less) can result in states with very long natural lifetimes.

Lifetimes and Quantum Yields

Since τ_o corresponds to the lifetime of an excited state when emission from that state is the only significant mode of decay, τ_o for the lowest excited singlet state S_1 will correspond to the observed lifetime only when the quantum yield of fluorescence φ_f is unity. Since φ_f is usually less than unity, the observed lifetime τ will be shorter than τ_o, that is

$$\tau = \varphi_f \tau_o \tag{5}$$

τ is the time required for a given population of excited states to decay to $1/e$ of the original population. Table I gives representative lifetimes for a wide variety of excited states.

It is perhaps easier to visualize the meaning of τ if we examine the kinetics of the processes which can cause a singlet excited state S_1 to decay following termination of excitation. These processes may be listed as follows:

$$S_1 \xrightarrow{k_f} S_o + h\nu \quad \text{(fluorescence)}$$

$$S_1 \xrightarrow{k_{ST}} T_1 \quad \text{(intersystem crossing)}$$

$$S_1 \xrightarrow{k^S_{IC}} S_o + \text{heat} \quad \text{(internal conversion)}$$

$$S_1 \xrightarrow{k^S_c} \text{products (unimolecular decomposition from } S_1)$$

$$S_1 + Q_1 \xrightarrow{k^S_{Q_1}[Q_1]} S_o + Q_1 + \text{heat (quenching)}$$

TABLE I

Lifetimes of Some Representative Excited States

System	Process	Lifetime (sec)
vibrationally excited primary excited state	$S_1(v_2) \rightarrow S_1(v_1)$ by collision induced de-excitation	$10^{-11} - 10^{-12}$
higher excited singlet state	$S_n \rightarrow S_1$	$10^{-12} - 10^{-14}$
lowest excited singlet state	Fluorescence, internal conversion, intersystem crossing	$10^{-7} - 10^{-9}$
fluorescein		$(4.5 \pm 0.4) \times 10^{-9}$
N-methylacridinium chloride		$(3.5 \pm 0.1) \times 10^{-8}$
lowest excited triplet $(n\pi^*)$ state	Phosphorescence, intersystem crossing	$10^{-2} - 10^{-4}$
m-iodobenzaldehyde		6.5×10^{-4}
benzophenone		4.7×10^{-3}
lowest excited triplet $(\pi\pi^*)$ state	Phosphorescence, intersystem crossing	$10^{+1} - 10^{-2}$
1-bromonaphthalene		1.8×10^{-2}
naphthalene		2.3
perdeuteronaphthalene		9.5

$$S_1 + Q_1 \xrightarrow{k^S_{Q_2}[Q_2]} S_o + Q_2^* \quad \text{(energy transfer)}$$

$$S_1 + Q_3 \xrightarrow{k^S_{Q_3}[Q_3]} \text{products (bimolecular chemical reaction from } S_1)$$

Suppose that the light is cut off at t=0 and that a certain concentration $[S_1]_o$ of S_1 exists in the sample. If no more light is absorbed by the sample, the rate of disappearance of S_1 will be given by

$$\frac{-d[S_1]}{dt} = (\sum_i k_i^S) \, [S_1] \tag{6}$$

where $\sum_i k_i^S = k_f + k_{ST} + k_{IC}^S + k_c^S + \sum_i k_{Q_i}^S [Q_i]$ (7)

Here $\sum_i k_{Q_i}^S [Q_i]$ extends over all bimolecular decay processes from S_1. Note that the bimolecular processes are pseudo-first-order, i.e. the $[Q_i]$ are constant during the decay of S_1. Equation (6) is a first-order differential equation for which the solution is

$$[S_1] = [S_1]_o \exp[-(\sum_i k_i^S)t] \tag{8}$$

All the terms in Equation (7) are first-order rate constants. Hence, we may define a time τ such that

$$\tau = (\sum_i k_i^S)^{-1} \tag{9}$$

or Equation (8) becomes

$$[S_1] = [S_1]_o \exp[-t/\tau] \tag{10}$$

The quantum yield of fluorescence φ_f is defined as

$$\varphi_f = \frac{\text{rate of fluorescence emission}}{\text{rate of absorption of light}} \qquad (11)$$

Under steady-state conditions, the rate of absorption into S_1 must equal the total rate of decay by all of the above processes. Hence,

$$\varphi_f = \frac{k_f [S_1]}{\sum_i k_i^s [S_1]} = \frac{k_f}{\sum_i k_i^s} \qquad (12)$$

Since $\tau_o = k_f^{-1}$ then

$$\varphi_f = \frac{\tau}{\tau_o}$$

or $\quad \tau = \varphi_f \tau_o$ $\qquad (5)$

Another important quantum yield is that of phosphorescence, φ_p, defined as

$$\varphi_p = \frac{\text{rate of phosphorescence emission}}{\text{rate of absorption of light}} \qquad (13)$$

By analogy with Equation (12) we may write

$$\varphi_p = \frac{k_p [T_1]}{\sum_i k_i^s [S_1]} \qquad (14)$$

since at steady state the rate of absorption must equal the

101

rate of decay of S_1 [see Equation (6)]. Also at steady state the rates of formation and decay of T_1 must be equal or

$$k_{ST}[S_1] \ = \ \sum_j k_j^t [T_1] \tag{15}$$

where

$$\sum_j k_j^t = k_p^{\cdot} + k_{IC}^t + k_c^t + \sum_i k_{Q_i}^t [Q_i] \tag{16}$$

Thus from Equation (15)

$$\frac{[T_1]}{[S_1]} = \frac{k_{ST}}{\sum_i k_j^t} \tag{17}$$

and by substituting Equation (16) into Equation (14)

$$\varphi_p = \frac{k_p k_{ST}}{\sum_i k_i^s \sum_j k_j^t} \tag{18}$$

Another way of looking at Equation (18) is that

$$\begin{pmatrix} \text{probability of} \\ \text{phosphorescence} \end{pmatrix} = \begin{pmatrix} \text{probability for} \\ \text{intersystem crossing} \end{pmatrix} \begin{pmatrix} \text{probability of} \\ T_1 \text{ emitting} \end{pmatrix}$$

$$\text{or} \quad \varphi_p = (k_{ST} \tau_f) (k_p \tau_p) \tag{19}$$

$$\varphi_p \ = \ \frac{k_{ST}}{\sum\limits_{i} k_i^s} \cdot \frac{k_p}{\sum\limits_{j} k_j^t} \qquad (18)$$

Equation (18) can be simplified by noting that

$$\varphi_f \ = \ \frac{k_f}{\sum\limits_{i} k_i^s} \qquad (12)$$

Hence

$$\frac{\varphi_p}{\varphi_f} \ = \ \frac{k_p k_{ST}}{k_f \sum\limits_{j} k_j^t} \qquad (20)$$

For some molecules phosphorescence is the only significant mode of decay from T_1 and intersystem crossing plus fluorescence are the only significant processes by which S_1 decays. In such cases, $\sum\limits_{j} k_j^t \simeq k_p$ and Equation (20) simplifies to

$$\frac{\varphi_p}{\varphi_f} \ = \ \frac{k_{ST}}{k_f} \qquad (21)$$

k_f can be estimated from the absorption spectrum using Equation (2) and thus in this special case k_{ST} can be determined. However, in general k_{ST} is difficult to obtain. Note that k_{ST} cannot be obtained by simply measuring the decay of S_1, since from Equations (6) and (7), the decay rate constant depends on other factors.

103

It is of some interest to use the equations developed above to calculate the concentrations of S_1 and T_1 under various conditions. These concentrations depend on the nature and intensity of the exciting light source and on the various rate constatns involved. Several cases will be considered.

Steady-state Irradiation

In each of the steady-state calculations below we will use a typical medium pressure mercury lamp producing a collimated beam with an intensity of 20 watts cm^{-2} at ~ 250 nm. We will assume that half the incident light is absorbed (i.e. sample O.D. ~ 0.3). The incident light intensity must be expressed in Einsteins cm^{-2} sec^{-1} (1 Einstein = 1 mole of photons), i.e. the absorbed light intensity $I_a = I_0/2 = 2.1 \times 10^{-5}$ Einsteins cm^{-2} sec^{-1}. We will assume a sample volume of 1 cm^3.

Case 1: $k_{ST} \ll k_f$; $\varphi_f \sim 1.0$; $k_f \sim 10^8$ sec^{-1}

In this case almost all of S_1 decays by fluorescence. At steady state, the rate of change of S_1 must be zero or

$$\frac{d[S_1]}{dt} = I_a - k_f[S_1] = 0 \tag{22}$$

or

$$[S_1] = \frac{I_a \times 10^3}{k_f} = 2.1 \times 10^{-10} \text{moles } \ell^{-1} \tag{23}$$

Case 2:

$$k_{ST} \gg k_f; \quad \varphi_p \sim 1.0; \quad k_p \sim 1$$

Here S_1 converts to T_1 so rapidly that fluorescence can be ignored. Since $\varphi_p \sim 1.0$, we can assume that phosphorescence is the only significant mode of decay of T_1.

104

Again at steady state the rate of change of T_1 must be zero. Since $k_{ST} \gg k_f$, the rate or production of T_1 is I_a. Thus

$$\frac{d[T_1]}{dt} = I_a - k_p[T_1] = 0 \qquad (24)$$

or

$$[T_1] = \frac{I_a \times 10^3}{k_p} = 2.1 \times 10^{-2} \text{moles } \ell^{-1} \qquad (25)$$

Note that for the same incident light intensity the steady state concentration of T_1 in Case 2 is much higher than that of S_1 in Case 1.

Note that in this case we have assumed that the only mode of decay of T_1 is by phosphorescence. In a low temperature glass or the solid state this situation is approached in some cases; however, for liquid solutions bimolecular quenching processes generally predominate and determine the rate of triplet decay. This will be discussed later.

Flash Illumination

In a transient photolysis situation it is convenient to assume that all the light from the flash source is emitted in a negligibly short period of time (a so-called δ-pulse). With the advent of laser flash sources with lifetimes 1 nsec. or less, this approximation is usually quite good; however, with conventional flash lamps, one must be concerned with the lifetime and intensity profile of the flash.

For the purposes of these illustrative calculations, we will take a frequency doubled ruby laser (347.2 nm) with a total energy per flash of 0.6J. Of this we will assume that 0.3J is absorbed by the sample. This corresponds to the absorption of $\sim 10^{-6}$ einsteins. Since each photon absorbed should excite one molecule to the S_1 state, the

105

initial yield of S_1 will be $\sim 1\times10^{-5}$ moles. If the sample is contained in a cell of 1 cm^3 volume, then the initial concentration of S_1 will be 10^{-3} moles ℓ^{-1}. Thus, quite high concentrations of excited states can be obtained with flash sources. If the width of the flash is much greater than the fluorescence decay time, then $[S_1]$ follows the intensity profile of the lamp, and calculations become more complicated.

Photochemical Reactions

An interesting application of the kinetics developed above is found in the cis-trans photoisomerization of olefins

In general, the absorption spectra of the two isomers will be different and hence at any given wavelength one or the other isomer will have a higher extinction coefficient. Usually at longer wavelengths $\epsilon_t > \epsilon_c$ where ϵ_t and ϵ_c are the extinction coefficients for trans and cis isomers respectively. Let φ_{tc} and φ_{ct} be the quantum yields for trans → cis and cis → trans reactions respectively.

First we will consider only light absorbed by the cis isomer. The rate of formation of the trans isomer by this process will be

$$\frac{d[\text{trans}]}{dt} = k_{ct}[\text{cis}] \cong \varphi_{ct}\epsilon_c I_o[\text{cis}] \tag{26}$$

Similarly for the trans → cis conversion

$$\frac{d[\text{cis}]}{dt} = k_{tc}[\text{trans}] \cong \varphi_{tc}\epsilon_t I_o[\text{trans}] \tag{27}$$

Equation (26) is valid only when the optical density is small. Then one may use the approximation $\dfrac{I_a}{I_o} \sim 0.D. =$ $\epsilon c \ell$ where ϵ is the extinction coefficient, c is the concentration of the absorbing species and ℓ is the path-length. I_o is in units of einsteins $cm^{-2}sec^{-1}$.

At steady state these two rates must be equal or

$$\frac{[cis]}{[trans]} \cong \frac{\varphi_{tc}\epsilon_t}{\varphi_{ct}\epsilon_c} \tag{28}$$

For many olefins at thermal equilibrium $[cis]/[trans]$ $\ll 1$. However, since ϵ_t/ϵ_c is often greater than one and φ_{tc} and φ_{ct} are usually comparable, a photostationary state will have $[cis]/[trans] > 1$.

Quenching

Here we shall use quenching in the broad sense as any bimolecular reaction between an excited-state molecule (S_1 or T_1) and a quencher molecule Q which results in the disappearance of S_1 or T_1. The effect of the quencher will thus be to decrease the quantum yield of fluorescence or of phosphorescence. When bimolecular quenching is absent (i.e. the concentrations of all possible quenchers are zero), the quantum yield for fluorescence is given by

$$\varphi_f^o = \frac{k_f}{k_f + k_{ST} + k_{IC}^s} \tag{29}$$

When quenchers are present

$$\varphi_f = \frac{k_f}{k_f + k_{ST} + k_{IC}^s + \sum\limits_i k_{Q_i}^s [Q_i]} \tag{30}$$

Now
$$\frac{\varphi_f^\circ}{\varphi_f} = 1 + \frac{\sum\limits_i k_{Q_i}^s [Q_i]}{k_f + k_{ST} + k_{IC}^s} \tag{31}$$

But from Equation (9) the lifetime in the absence of quenchers is

$$\tau_f = \frac{1}{k_f + k_{ST} + k_{IC}^s}$$

Thus
$$\frac{\varphi_f^\circ}{\varphi_f} = 1 + \tau_f \sum\limits_i k_{Q_i}^s [Q_i] \tag{32}$$

If only one quencher is important then

$$\frac{\varphi_f^\circ}{\varphi_f} = 1 + \tau_f k_Q^s [Q] \tag{33}$$

Since the intensity of fluorescence I_f will be proportional to φ_f, Equation (33) can be written as

$$\frac{I_f^\circ}{I_f} = 1 + \tau_f k_Q^s [Q] \tag{34}$$

Equation (34) is known as a Stern-Volmer relation. A similar relation applies to phosphorescence. It should be noted that if more than one quencher is important, the Stern-Volmer plot of $\varphi_f^\circ/\varphi_f$ vs [Q] may become non-linear.

The slope of a Stern-Volmer plot gives the product $\tau_f k_Q^s$. Clearly, additional information is necessary to separate these two factors. In some cases τ_f can be determined directly by observing the decay of fluorescence. Alternatively τ_f may be estimated using Equations (2) and (5) and a measurement of φ_f.

It is often found that k_Q^s approaches the diffusion controlled limit in solution. This is the rate which would be obtained if every encounter of the excited state with a quencher molecule resulted in quenching. An approximate expression for the diffusion controlled rate constant k_{dif} is given by the Debye equation

$$k_{dif} = \frac{8RT}{3000\eta} \quad \ell \text{ mole}^{-1} \text{ sec}^{-1} \tag{35}$$

where η is the viscosity in poise. For aqueous solutions at $\sim 20^\circ C$, $k_{dif} \sim 7 \times 10^9 \; \ell \text{ mole}^{-1} \text{ sec}^{-1}$. For cyclohexane solutions, it is about $1 \times 10^{10} \; \ell \text{ mole}^{-1} \text{ sec}^{-1}$.

The quencher concentration $[Q]_{1/2}$ is defined as that concentration of Q which will reduce the fluorescence intensity to half of the value with no quencher present. From Equation (34) one can derive that

$$[Q]_{1/2} = \frac{1}{k_Q^s \tau_f} \tag{36}$$

If k_Q^s is diffusion controlled (i.e. $\sim 10^{10} \; \ell \text{ mole}^{-1} \text{ sec}^{-1}$) and $\tau_f \sim 10^{-8}$ sec then typically $[Q]_{1/2} \sim 10^{-2}M$ for the quenching of fluorescence. Clearly, concentrations of Q less than $10^{-3}M$ will have little effect on fluorescence, especially if k_Q^s is less than k_{dif}. However, for triplet states τ_p can be 10^{-3} - 10 sec. For $\tau_p \sim 1$ sec and $k_Q^s \sim 10^{10} \; \ell \text{ mole}^{-1} \text{ sec}^{-1}$ $[Q]_{1/2} \sim 10^{-10}M$, and trace quantities of impurities, if effective as quenchers, can completely remove phosphorescence in solution. In fact, the triplet states may undergo self-quenching through the process of triplet-triplet annihilation

$$T_1 + T_1 \rightarrow S_0 + S_1$$

From this discussion of quenching kinetics it is clear that either impurity quenchers or triplet-triplet annihilation are effective in removing T_1 and indeed phosphorescence is rarely observed in liquid solutions. Of course, in crystals or in low temperature glasses diffusion is very slow and hence quenching is not effective.

A study of half-quencher concentrations can be useful in characterizing an excited state. If the quencher is effective below $\sim 10^{-4}$M, it is highly unlikely that the excited state is a singlet.

GENERAL REFERENCES

More detailed accounts of photochemical kinetics can be found in the following:

[1] J.B. Birks, "Photophysics of Aromatic Molecules", Wiley Interscience, New York, 1970.

[2] J.G. Calvert and J.N. Pitts, Jr., "Photochemistry", John Wiley and Sons, New York, 1966, p. 174.

[3] W.R. Ware in "Creation and Detection of the Excited State", Ed. by A.A. Lamola, Marcel Dekker Inc., New York, 1971, Vol. 1, Part A.

CHAPTER 5
EXPERIMENTAL TECHNIQUES

In this Chapter we will be concerned with a very brief description of light sources, techniques for measuring light intensities and quantum yields, methods used to detect transient photochemical products and some ancilliary techniques.

Steady-State Light Sources

One need not go to the expense of purchasing a light source for photochemical experiments as the sun can and has been used for this purpose. However, if one wishes to be free of problems such as clouds and the rotation of the earth, then an artificial source is needed. Since most photochemical reactions require ultraviolet light, attention has naturally focused on effective sources in that region. The most popular sources by far are mercury arcs which are available in three types: the low pressure (or resonance) lamp (pressure $\sim 10^{-3}$ torr); the medium pressure arc (pressure ~ 35 torr) and the high pressure arc (pressure ~ 100 at.).

In the low pressure mercury resonance lamp most of the radiation ($\sim 90\%$) is concentrated at 253.7 nm. This lamp is an excellent source for mercury photosensitized reactions; however, the intrinsic brilliance (intensity per unit area) of the lamp is low, and thus its use in photochemical syntheses is limited.

The medium pressure lamp has a much broader spectral distribution than the low pressure lamp as many more emission lines appear. A typical emission spectrum is given in Figure 1a. This lamp is well suited to a wide range of applications.

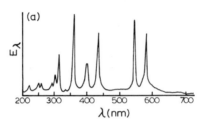

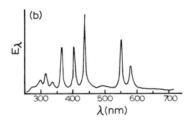

Figure 1.   a)   Medium pressure lamp
             b)   High pressure lamp

The high pressure mercury lamp is used in applications where a very high brilliance and a small source area are required (Figure 1b shows the emission spectrum of such a lamp). This type of lamp is used in applications where the sample to be irradiated is small, such as a sample in an electron spin resonance cavity (see below), or where the slits of a monochromator are to be illuminated.

Many other types of lamps may be used including rare-gas discharge lamps. For instance, the xenon lamp has a spectral distribution which is very similar to that of the sun (see Figure 2) and is useful in simulating solar effects. This lamp also has a strong and continuous output in the visible region and is used where high intensities in that region are required. It also has useful intensity in the ultraviolet.

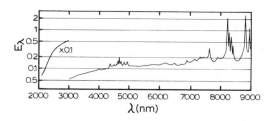

Figure 2.   Xenon lamp spectrum.

In situations where both ultraviolet and visible radiation is required, combination mercury-xenon lamps are available (see Figure 3 for the spectral distribution of this type of lamp).

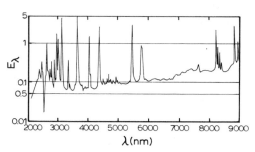

Figure 3.   Mercury xenon lamp spectrum.

Some of the lamps described above can be operated with either an AC or DC. A series resistance or current regulation cicuit is required to control the current through the lamp, especially in the warm-up phase. For quantitative

113

work where a constant intensity is required, DC operation is imperative. Also, many lamps are designed to operate only on DC.

Flash Lamps

The availability of high intensity, short duration flash lamps has been a great boon to the photochemist in the past several decades. Most of the sources used have been gas discharge tubes consisting of two electrodes encased in a tube filled with a moderate pressure (6 - 20 cm) of a gas such as xenon. The electrical energy ($1/2\ CV^2$) for each flash discharge is stored in a bank of condensers (of capacitance C farads) charged to a high voltage (V). Giant flashlamp systems have in fact been built in which as much as 15 - 20 kJ of electrical energy is dissipated per flash. In one such instrument the condenser bank occupies an entire room and the current is led to the flash room via large copper bars. More conventional systems employ 100J to a few kJ.

Two methods are used to initiate the flash. In one the high voltage applied to the lamp is kept well below the spontaneous discharge point (see Figure 4). The lamp is flashed by applying a very high voltage (15 - 20 kV) pulse to a small trigger wire wound around the flash tube. This trigger pulse ionizes some of the gas molecules and thus initiates breakdown.

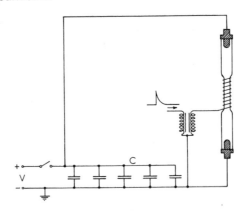

Figure 4.   Flash lamp circuit.

In the second method, the high voltage is set well above the spontaneous discharge point of the flashlamp but is held off by a spark gap (see Figure 5). When the gap is sparked by a high voltage trigger pulse, the current from the capacitor bank discharges through the flashlamp. This latter method is preferable when it is desired to keep the flash as short as possible since much higher voltages may be used.

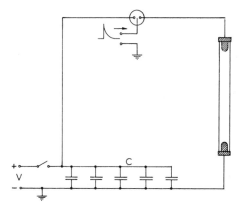

Figure 5.   Spark-gap triggered lamp.

In a typical flashlamp about 2 - 5% of the electrical energy is converted into light energy. From a knowledge of the spectral distribution of the flash and the total light output one can calculate the number of Einsteins of light in a given wavelength region from the following equation:

No. of Einsteins = total light energy in the wavelength region of interest

energy per photon x N_o

$$= \frac{E(\lambda)\lambda}{hcN_o} = 8.359 \times 10^{-9}\lambda E(\lambda) \quad (1)$$

where N_o is Avogadro's number, $E(\lambda)$ is the energy in joules in the wavelength region of interest and λ, the average wavelength of the light, in nm.

For a typical flash photolysis system involving electrical energies of the order of 4000 - 5000 joules per flash, approximately 10^{-2} Einsteins at ~ 300 nm can be delivered to a sample tube. In most cases this is sufficient to cause photolysis of a large percentage of the sample, or the conversion of a large fraction of the sample to its triplet state.

There is a definite relation between the flash duration (pulse width at half-maximum) and the energy of the flash. Flash times can be minimized through proper design, but for most flash photolysis systems involving discharges in the 1000 to 5000 J range, flash times below 1 msec. are difficult to obtain. Flashlamps in the ~ 100 J range can be made with flashtimes of a few μsec. Some flashlamps are available with decay times ~ 1 - 5 nsec. but they dissipate nanojoules of electrical energy. While these lamps may be used in the study of photophysical processes in the nsec. time range, detection of the excited states must be by fluorescence techniques because of the low photon flux (10^6 - 10^{10} photons/flash).

Lasers

Laser stands for light amplification by stimulated emission of radiation, a technique which was suggested in 1958 by Schawlow and Townes at Columbia University, and first reduced to practice by Maiman at Hughes Research Laboratories. No attempt will be made here to explain the theory of laser action[1,2] except to comment that a laser

---

1. See "Lasers and Light" readings from Scientific American, N.H. Freeman, 1969. Secs. VI and VII contain excellent elementary discussion of the theory and application of lasers.

2. "Lasers" B.A. Lengyel, Wiley Interscience, New York, Second Edition, 1971.

116

usually produces a highly collimated, monochromatic and coherent beam of light. Both pulsed and continuous lasers have been developed and are commercially available.

The great advantage to a photochemist provided by lasers can be appreciated by reference to the earlier discussion of flashlamps. Many photochemical processes occur within a few nanoseconds. However, conventional flashlamps with lifetimes in the nanosecond region put out a very low light flux. Giant Q-switched ruby lasers have been built which have an output of 5 - 10 joules or more at 694.3 nm. Use of a non-linear cyrstal permits doubling of the light frequency ($\lambda = 347.2$ nm) with an output of 1 joule or more or ~ 3×10^{-6} Einsteins. This is certainly sufficient to photolyze a large fraction of a sample. However, the striking advantage is that the pulse from the giant pulsed ruby laser lasts only 20 nsec. with very little tail. With the advent of Q-switched lasers, the state of the art in flash photolysis jumped from μsec. to nsec. Unfortunately the doubled ruby laser is not tunable and 347.2 nm is not far enough into the ultraviolet to reach the absorption bands of many compounds of interest to photochemists. There is, however, another high powered solid-state laser using neodymium glass rod that lases at 1.06 microns. Light at this frequency can be doubled twice (quadrupled) to yield a 15 - 20 nsec. pulse of 0.02 - 0.05J at 265 nm. Both the photon flux per pulse and the wavelength are well suited for laser flash photolysis of many classes of compounds.

CW (continuous) ion lasers are also useful as a photolysis source. Photon fluxes of 10^{15} to 10^{18} photons per second are available in a variety of lines from 360 - 700 nm. These lasers utilize argon or krypton as the lasing gas. Cadmium ion CW lasers operating at 325 and 436 nm are also useful. CW lasers can also be employed for direct production of triplets by irradiation in the S - T band, and as sources for fluorescence and phosphorescence spectroscopy.

The lasers described above operate at fixed wavelengths. A variety of tunable lasers also are available which in general make use of a dye solution placed in a laser cavity. The dye molecules are the lasing species and tuning is accomplished over a portion of the dye emission band with a prism or grating inside the cavity. Wavelengths from 340 nm to 700 nm are available from dye lasers with pulse energies as high at 1 - 2J per pulse and durations of several hundred nanoseconds. Dye lasers of lower power pumped with the nitrogen superradiant laser are also available, which typically have a pulse duration of 7 - 10 nsec. and cover a range of 360 - 500 nm. With tunable dye lasers bandwidths of 1Å are typical and 0.001Å is not difficult to achieve.

The field of laser technology is moving very rapidly and one can anticipate remarkable changes in laser light sources in the near future which will provide faster and more intense pulses and tunability well below 300 nm with a very narrow band width. Conventional light sources such as xenon and mercury lamps may well be completely obsolete within a decade or less.

Measurement of Quantum Yields

In the determination of a quantum yield two kinds of measurements are required. The first is a determination of the amount of a product; the second is the determination of the amount of light absorbed. That is

$$\Phi_x = \frac{\text{No. of molecules undergoing process X}}{\text{No. of absorbed protons}} \tag{2}$$

Both will be commented on separately.

The "product" of photoexcitation need not necessarily be a chemical product; that is, the term quantum yield may be applied to any measurable phenomenon and includes phosphorescence or fluorescence. Absolute measurements of light intensity of emission are amongst the most difficult to make.

Where the products are chemical, analytical means for estimation of product can be employed, but the quantum yield arrived at will be no better than the analytical tool. To avoid secondary reactions, only very small photochemical conversions should be employed and it is clear that very sensitive detection systems are required. Of these vapor phase chromatography has probably become the most important. An elementary precaution should always be to ascertain that the quantum yield is independent of the degree of conversion (over the range studied) and that it is likewise independent of the light intensity: i.e. that there are no biphotonic processes. This latter precaution will become very important when high intensity lasers are used for excitation.

When the quantum yield is very low the amount of product available for analysis is limited by the degree of conversion possible in reasonable time, say, 20 - 40 hours. Under such circumstances ($\Phi < 10^{-4}$) special techniques must be employed. Isotopic dilution has been used and this elegant technique may be the best. However, when it is recollected that several determinations may be necessary for a Stern-Volmer plot, this procedure is enormously time-consuming and a compromise may have to be reached. For the most part the analytical techniques are standard, and outside the scope of this book, but perhaps a warning should be given against, where possible, employing the often simpler procedure of measuring the disappearance of starting material. Evidently the quantum yield of disappearance of starting material and appearance of product will be the same if there is only one product, and detection of only one product does not rule out the presence of others. In addition, with low conversions there are difficulties in measuring the difference between two large figures. If conversions are carried further, then the problem of secondary chemical or photochemical transformations may become important. The references appended to this section give examples of the use of a number of analytical techniques.

To obtain a quantum yield the product yield has to be related to the light absorbed. There are a number of considerations involved here. First, the flux incident on the device - usually a cell - in which the reaction is being performed may be measured electrically using some detector system, or it may be measured taking advantage of the known quantum yield for a well-studied reaction. Secondly, one may measure the flux before and after the ex-

periment and assume the flux has been the average during
the experiment, or one may use a beam-splitter - frequently
a plate of quartz at 45° to the beam - where a constant
pre-determined proportion of the light is deflected and
measured while the reaction is in progress. No assumptions
about intensity fluctuation are then necessary. The measurement is most simple if all the incident light is absorbed
by the medium, but it is desirable that it should not all
be absorbed in too small a fraction of the total pathway
since problems of diffusion and local over-irradiation may
become very serious. If all the incident light is not absorbed then either a detection system behind the reaction
cell is required, or else one must calculate the amount of
light absorbed from a knowledge of the optical density over
the wavelength range of irradiation.

Most frequently chemical actinometers are used to
measure the amount of light absorbed. Those preferred will
depend on the wavelengths of interest. Some have quantum
yields that are quite variable with wavelength and hence
are suitable only for use with monochromatic light. Others,
and of these the most used is the ferrioxalate actinometer
of Parker and Hatchard, have values that are more or less
constant over a broad band. The latter may be used from
about 250 nm to 450 nm. It must be remembered that every
glass or quartz surface introduced into the light beam in-
troduces errors due to reflection (internally and externally
in the case of a cell).

Apparatus

The apparatus used for chemical quantum yields may be
of varying degree of complexity. The classical optical
bench is the most common and allows the insertion in a
collimated light beam of a monochromator or filters as
desired. A likely source of error here may be scattered
light of a wavelength supposedly suppressed. More sophisti-
cated devices have been evolved for multiple irradiation.
In these there is a fixed light source, either central or
on the periphery of a circle. A series of tubes is rotated
within that circle. Under these conditions it is assumed
that all tubes are equally irradiated. If provision is
made for filters and some tubes contain actinometer solu-

tion, quantitative results can be obtained.

The danger of systematic error in rotating these systems, if the tubes are totally exposed to the light, is very great if for no other reason than at the edge of the tube the pathlength is very short. If the concentration is so high that total absorption occurs here, then there may be a question of local over-irradiation. In such apparatus there is no way of measuring transmitted light. Enclosure of the tubes permitting irradiation of only the center of the tube is far superior and also reduces internal reflections. In principle, it should be possible in such systems to measure transmitted light.

Electrical Measurement of Light Intensities

Light intensities may be quoted either in units of energy sec^{-1} cm^{-2} (watts cm^{-2} or erg sec^{-1} cm^{-2}) or in units of einsteins sec^{-1} cm^{-2}. The various conversion factors are as follows:

$$1 \text{ einstein } sec^{-1} \ cm^{-2} = \frac{1.196 \times 10^{15} \text{ erg } sec^{-1} \ cm^{-1}}{\lambda}$$

$$= \frac{1.196 \times 10^{8} \text{ watt } cm^{-2}}{\lambda} \tag{3}$$

$$1 \text{ watt } cm^{-2} = 10^{7} \text{ erg } sec^{-1} \ cm^{-2} =$$
$$8.358 \times 10^{-9} \ \lambda \text{ Einstein } sec^{-1} \ cm^{-2} \tag{4}$$

A useful conversion table is given below.

A reliable electrical method of measuring light intensities is to use a thermopile. This device contains a black element which totally absorbs all incident radiation, independent of wavelength. The resulting temperature rise in the element is detected by a number of thermocouples connected in series; an equal number of thermocouples not exposed to the radiation provide a reference voltage for bridge circuit. The output of the bridge is usually displayed on a chart recorder of microvoltmeter and measurements are made with a fixed light on - light off cycle.

121

TABLE 1   Energy Conversion Factors

To get → Multiply By ↓	METER-CANDLES = LUX = $\dfrac{\text{LUMENS}}{m^2}$	FT-CANDLES = $\dfrac{\text{LUMENS}}{ft^2}$	PHOT = $\dfrac{\text{LUMENS}}{cm^2}$	$\dfrac{\text{WATTS}}{ft^2}$	$\dfrac{\text{WATTS}}{m^2}$	$\dfrac{\text{WATTS}}{cm^2}$
METER-CANDLES = LUX = $\dfrac{\text{LUMENS}}{m^2}$	1	9.29×10^{-2}	10^{-4}	1.50×10^{-4}	1.61×10^{-3}	1.61×10^{-7}
FT-CANDLES = LUMENS/ft^2	10.76	1	10.76×10^{-4}	1.61×10^{-3}	1.73×10^{-2}	1.73×10^{-6}
PHOT = LUMENS/cm^2	10^4	9.29×10^2	1	1.50	16.1	1.61×10^{-3}
WATTS/ft^2	6.67×10^3	6.21×10^2	0.667	1	10.76	10.76×10^{-4}
WATTS/m^2	6.21×10^2	57.7	6.21×10^{-2}	9.29×10^{-2}	1	10^{-4}
WATTS/cm^2	6.21×10^6	5.77×10^5	6.21×10^2	9.29×10^2	10^4	1

When standardized against a National Bureau of Standards
Certified standard lamp, a thermopile can measure absolute
light intensities to ~ 1 - 2%. However, the device is
rather insensitive and since it relies on integrating the
light over a certain period of time, its principal utility
is where the light to be measured is relatively constant in
time.

A more sensitive instrument, which also works on the
energy absorbing principle, is the radiometer. A radiometer uses a thermistor in place of thermocouples and thus
achieves much higher sensitivity. Incident intensities as
low as 1 mW cm^{-2} can be measured with ease. It should be
noted that both the thermopile and the radiometer measure
incident power irrespective of the wavelength of the light.

Both the radiometer and the thermopile are useful for
measuring intensities of continuous soucres, although in-
termittent sources can be measured if the repetition rate
is high (> 10 flashes per second). With flash sources,
one is interested in the total light energy per flash. Here
one can use a light calorimeter which totally absorbs the
light, a technique which is especially useful for laser
sources. Energies as low as 10 mJ can be measured.

The detection of light intensities at lower levels requires more sensitive techniques. Here the use of a photocell or the much more sensitive photomultiplier is ad-
vantageous. They function as follows. A light-sensitive
cathode in an evacuated tube liberates electrons when
irradiated with light (this is the famous photoelectric
effect). The electrons are collected at a single anode in
the photocell; in the photomultiplier the current of electrons is amplified by the use of a number of dynodes. In
each device the current produced is normally proportional
to the light intensity; both devices have a sensitivity
which varies with wavelength and thus they are not very
satisfactory for absolute intensity measurements. However
they may be calibrated against a thermopile or radiometer.
For relative measurements they are extremely sensitive.
(The typical photomultiplier has a gain of 10^7.) In fact
photomultipliers are available which permit one to count
single photons and produce up to a volt pulse amplitude per

photon into 50 ohms. Another advantage of photomultipliers is that they respond very rapidly when the light intensity changes. Response times of 2 - 3 nsec. are typical.

Emission Quantum Yields

The measurement of emission quantum yields is much more difficult. The product is now the light emitted, generally over the total 4π solid angle, and one must know the fraction of the total emitted photon flux that is actually measured as well as the number of photons absorbed in the measuring instrument. Relative measurements against so-called standards are frequently made using $90°$ geometry in spectrofluorimeters. One measures the fluorescence spectrum of the standard and the unknown, corrects for the spectral response of the detector system (monochromator and photomultiplier) and calculates the quantum yield with the assumption that the standard gives the geometrical factor. Account must also be taken of differences in the amount of light absorbed by the standard and the unknown. A more satisfactory approach involves the use of an integrating sphere which, through multiple reflections, eliminates arti- facts due to the index of refraction variation between air, glass and the solution. However, one ultimately must have a standard, and this implies absolute measurements. Ab- solute fluorescence quantum yields standards are at present accurate to no better than 5 - 10% and relative measure- ments may be subject to large systematic errors.

The true yield of phosphorescence depends on both the efficiency of intersystem crossing and the efficiency with which the triplet emits once it is formed. If one can assume that $\varphi_{isc} = 1 - \varphi_F$, where φ_{isc} is the intersystem crossing yield and φ_F the fluorescence quantum yield, then the fraction of the triplets that emit is given by the phosphorescence quantum yield times the quantum yield for intersystem crossing. However, it is not safe in general to assume that $\varphi_{isc} = 1 - \varphi_F$ and the true phosphorescence quantum yield is thus more difficult to determine than the fluorescence quantum yield. Nevertheless, it is an import- ant quantity and is frequently measured or approximated in connection with photophysical studies. There is no funda- mental difference in the experimental procedure for measuring phosphorescence quantum yields as compared to fluorescence quantum yields. Use of low temperature rigid glass matrices or host crystals eliminates quenching effects, but these effects can also be eliminated by working in plastics at room temperature, or by extremely rigorous purification of solvents.

References Regarding Quantum Yields

General

1) H.E. Johns in "Creation and Detection of the Excited State", A. Lamola, Editor, Marcel Dekker, Vol. 1, Part A, 1971, p. 123.

2) J.G. Calvert and J.N. Pitts in "Photochemistry", Wiley, New York, 1966.

3) H.E. Zimmerman, Mol. Photochem., 3, 281 (1971): of particular interest to organic chemists.

Rotating Quantum Yield Systems

F.G. Moses, R.S.H. Liu and B.M. Monroe, Mol. Photochem., 1, 245 (1969).

Actinometers

(Ferrioxalate)   C.G. Hatchard and C.A. Parker, Proc. Roy. Soc. (London), A235, 518 (1956).

(Malachite Green)   G.J. Fisher, J.C. LeBlanc and H.E. Johns, Photochem. Photobiol., 6, 757 (1967).

(Uranyl Oxalate)   C.A. Discher, P.F. Smith, I. Lippman and R. Turse, J. Phys. Chem., 67, 2501 (1963).

(Reinecke's Salt)   E.E. Wegner and A.W. Adamson, J. Amer. Chem. Soc., 88, 394 (1966).

Fluorescence Quantum Yields

J.N. Demas and G.A. Crosby, J. Phys. Chem., 75, 991 (1971).

Fluorescence and Phosphorescence Lifetimes

It was pointed out in earlier chapters that the fluorescence lifetime is an important parameter in photochemical and photophysical kinetics. Knowledge of the fluorescence lifetime permits one to determine second order quenching rate constants and to separate radiative from non-radiative transition probabilities of the excited singlet. Because of the short lifetime of this state (10^{-8} - 10^{-9} sec. is typical) special techniques are required for the determination of fluorescence lifetimes. The simplest approach is to use a nanosecond flash lamp and measure the fluorescence decay with a fast (e.g. pulse sampling) oscilloscope and a photomultiplier. This method unfortunately lacks sensitivity. Another approach is to employ a continuous light source which is intensity modulated at 1 - 20 MHz. The fluorescence then lags the exciting light by a certain phase angle, and the measurement of this angle allows one to calculate the lifetime of the excited singlet from an assumed decay law.

A very sensitive and currently popular approach involves the use of a nanosecond flash lamp coupled with photon counting and timing techniques. This technique is called the single photon method. Fluorescence photons are individually counted and the time of their observation is related to the time of excitation by an electronic device. The experiment is repeated at a very high repetition rate, typically 20 - 40 KHz, with fluorescence photons collected at 20 - 1000 Hz. Knowledge of the time of arrival of fluorescence photons relative to the time of discharge of the flash lamp provides one with the information necessary to construct a decay curve. This technique can be used to measure fluorescence lifetimes down to 0.3 nsec, and it has even been used in the microsecond range.

Phosphorescence lifetimes are easily measured with an oscilloscope and microsecond flash lamp. If the decay time is longer than a msec, a conventional camera shutter and a steady light source suffice. One can also employ a rotating sector and observe the decay on an oscilloscope during the dark phase. For more information, the reader is directed to the references listed below[3,4].

Methods Used to Detect Photochemical Transients

Photochemical transients include excited singlet or triplet states, free radicals (such as CH_3COHCH_3) and diamagnetic transients (such as singlet methylene). One way to detect free radical transients is by trapping techniques such as condensing gas phase intermediates in a cold finger immediately after photolysis in a flow[5] system. The intermediates may then be studied at will by a variety of spectroscopic techniques. An interesting variant on this method is the technique of spin trapping[6] which involves reaction of very reactive radicals (such as methyl or phenyl) with nitrones to form stable nitroxides

$$R\cdot + R_2C = \overset{\overset{\displaystyle O^-}{|}}{\underset{+}{N}} - R'' \longrightarrow RR_2C - \overset{\overset{\displaystyle O\cdot}{|}}{N} - R''$$

The reaction may be carried out both in gas and liquid phases; the radical $R\cdot$ is identified by the electron spin resonance spectrum of the nitroxide.

---

3. W.R. Ware in "Creation and Detection of the Excited State", A. Lamola, Editor, Marcel Dekker, Vol. 1, Part A, 1971, p. 213 - 302.

4. J. Birks in "Progress in Reaction Kinetics", Pergamon Press, New York, Vol. 4, 1967, p. 239.

5. A.M. Bass and H.P. Broida, Eds., "Formation and Trapping of Free Radicals", Academic Press, New York, 1960.

6. E.G. Janzen, Accounts Chem. Res., 4, 31 (1971).

When more than just qualitative information about the transients is required, that is, when concentration information, especially as a function of time, is required, then one must use sensitive detection techniques with a high degree of time resolution. By far the most popular techniques to fill this role have been optical absorption and emission spectroscopy. A typical flash photolysis system using optical absorption detection is illustrated in Figure 6. The entire absorption spectrum of a transient(s) may be recorded on a photographic plate following dispersion by a grating or prism, or a small range of wavelengths may be monitored with a photomultiplier preceded by a monochromator. The former system provides a 'snapshot' of the system

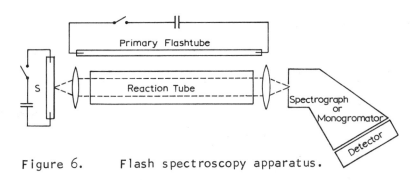

Figure 6.   Flash spectroscopy apparatus.

a set delay time after the flash while the latter system has the advantage of providing continuous kinetic information on a single transient species following the flash. If the transient emits light (e.g. an excited singlet or triplet), then the monitoring beam is not necessary; the detection system is then usually placed at right angles to the actinic light beam.

Optical absorption or emission detection has the advantage of sensitivity and fast time response. Often, however, optical bands of various species present may overlap making positive identification difficult and furthermore, one rarely has reliable extinction coefficients for the intermediates. Extinction coefficient measurements in fact imply knowledge of concentration. Thus one must use care in the analysis of results from this technique. Nevertheless, optical absorption or emission detection in conventional or laser flash photolysis has proved to be an invaluable tool in photochemical research.

Many of the transients produced in photochemical reactions are free radicals and electron spin resonance spectroscopy may be used to advantage because it is sensitive only to molecules containing unpaired electrons.

Electron spin resonance (esr) spectroscopy is based on the magnetic properties of unpaired electrons. In a magnetic field H, these electrons have energies given by

$$E = \pm 1/2 \, g\beta H$$

where g, which is approximately 2, is called the "g factor" and β is the Bohr magneton. If the system is now irradiated in the microwave frequency region (5 - 50 GHz), transitions may be induced between the energy levels provided the photon energy $h\nu$ matches the energy-level separation, that is

$$h\nu = g\beta H$$

This is the basic resonance relation of esr spectroscopy.

Two characteristics of esr spectroscopy should be pointed out. First, the spectra are almost always taken at fixed frequency and variable magnetic field; the resonant field H_r may vary from one molecule to another - this variation is characterized by the effective g factor

$$g = \frac{h\nu}{\beta H_r}$$

Second, the unpaired electrons may interact with magnetic nuclei in the molecule leading to a splitting of lines; this phenomenon is called hyperfine splitting.

Esr is a sensitive technique (radical concentrations as low as 10^{-9}M have been detected) and has proven to be a valuable adjunct to photochemical studies. If sufficient steady-state concentrations of radical intermediates can be generated by photolysis, the intermediates can be identified by their esr spectra. On the other hand, the coupling of flash photolysis techniques to esr spectroscopy allows detection of transients with lifetimes as short as a few μs. Unfortunately, the uncertainty principle prevents esr detection of transients with lifetimes less than ~ 1 μsec.

For further information beyond this very brief outline, the reader should consult some of the available references on esr[7,8].

Filters and Monochromators

Most photochemical studies require selection of certain wavelengths of light unless the light source itself is monochromatic. Monochromators are of two types: those based on a prism and those based on a grating. A prism monochromator has a high dispersion in the uv but the dispersion decreases to longer wavelengths. A grating monochromator scans linearly in wavelength as the grating is driven from a sine bar. This feature of the grating monochromator, plus its greater dispersion in the visible regions, has made it a popular choice.

Certain wavelengths may also be selected by means of filters, either chemical or glass. These are of two types: reflective and absorption filters.

The reflective filter operates on the interference principle. Interference filters can be obtained such that they will pass only a small bandwidth (5 - 20 nm) of light

7.   "Electron Spin Resonance, Elementary Theory and Practical Applications", J.E. Wertz and J.R. Bolton, McGraw-Hill, New York, 1972.

8.   J.R. Bolton and J.R. Warden, Jr. in "Creation and Detection of the Excited State", W.R. Ware, Editor, Marcel Dekker, Vol. 2, 1973 (in press).

about a selected center wavelength. All other wavelengths are reflected (higher orders are generally blocked with absorption filters). Transmission in the center of the band can be as high as 50%. These interference filters are valuable when moderately narrow band widths at high transmissions are required.

The absorption type of filter is usually a chemical either in solid or liquid solution which will absorb certain regions of the spectrum. A wide variety of colored glass filters are available. Also recipes are published for various combinations of solutions[9].

---

9.   J.G. Calvert and J.N. Pitts, Jr., "Photochemistry", John Wiley, New York, 1966, p. 729ff.

CHAPTER 6

PHOTOCHEMICAL PROCESSES
IN THE GAS PHASE

Introduction

Although the photochemistry of the gas phase and of
solutions is more noteworthy for similarities than for
differences, there is one important respect in which the
gas phase is less forgiving as a medium. Most photochemical transients are, at the instant of formation, endowed
with excess vibrational energy; indeed, some are produced
with excess electronic energy also, that is, in excited
electronic states. In solution, the excess energy is
rapidly removed by collision with solvent, so that the
transient reacts in its lowest vibrational and electronic
state. But in the vapor state, where the collision frequency is much lower - and dependent upon total pressure in
the system - the transient may not be energetically relaxed
before it reacts.

This Chapter divides into two parts: (i) a review of
some features of typical photochemical processes conducted
in the vapor state and (ii) a survey of the most chall-
enging and technically complex of all gaseous photochemical
systems, the reactions of the urban atmosphere.

1.   PHOTOSENSITIZATION

Photosensitization involves the absorption of radiation by a strongly absorbing substance, the photosensitizer
and its collisional transfer to another substance which is

non-absorbing at the same wavelength. Although it has considerable practical importance, for instance as a means of populating triplet states of molecules where direct excitation is usually not feasible, photosensitization is not merely an experimental convenience. The study of collisional energy transfer, in competition with spontaneous emission of radiation, is a problem of interest in its own right and presents a formidable challenge to theory and experiment.

Photosensitizers used in solution studies are usually large conjugated molecules with absorption in a convenient spectral region; benzophenone is a familiar example. Metal vapors have been particularly useful in gas phase work, though sulfur dioxide, benzene and other organic molecules are also employed. An advantage of the metal is that its atomic energy levels are well characterized and usually well separated, so that the energy available for energy transfer is sharply defined. A metal used for this purpose must, of course, meet the dual specification of a relatively high vapor pressure and an energy level in a photochemically-useful range (say, 50 - 120 kcal). These requirements narrow the choice to Na, Zn, Cd and Hg, of which the last has been the most extensively used. A low pressure arc of the same element serves as the external source of energy. For atoms of high atomic number, where the spin-orbit coupling is large, a transition from the ground state to excited triplet states has relatively high probability, thus the direct population of low-lying triplet states is easily accomplished (no change of multiplicity is involved in the excitation of Na atoms). Transitions populating these states are:

Metal	λ, nm	Transition	Energy, kcal
Na	589.6, 589.0	$^2S \rightarrow {}^2P^-$	58.4
Zn	307.6	$^1S_0 \rightarrow {}^3P_1$	92.5
Cd	326.1	$^1S_0 \rightarrow {}^3P_1$	87.3
Hg	253.7	$^1S_0 \rightarrow {}^3P_1$	112.2

Some low-lying energy levels of Hg are shown in Figure 1. Because spin-orbit coupling is large, the ^3P state of the atom is split into three distinct substates designated $^3P_0$, $^3P_1$ and $^3P_2$; the final subscript denotes the quantum number J of <u>total</u> angular momentum which is compounded from electronic orbital angular momentum L and electron spin angular momentum S, such that $J = L + S$. Since a ^3P state has $L = S - 1$, the possible quantum-mechanical values of J are 2, 1 and 0. Only the transition to the $J = 1$ sub-state, $^3P_1$, is allowed to occur through absorption of radiation by atoms in the ground state. $^3P_0$ lies 5 kcal lower in energy than $^3P_1$ and represents a metastable state which does not lose its energy by emission; consequently, its lifetime is of the order msec. The $^3P_1$ state is rapidly converted to $^3P_0$ by collisions with, e.g. N_2, CO or H_2O, but the reverse process requires only 5 kcal activation and also occurs readily. (Excitation from $^3P_1$ to $^3P_2$ requires about 12 kcal activation and is relatively unimportant.) The question whether mercury photo-sensitization involves $^3P_1$ or $^3P_0$ is complex and the two possibilities are not easily distinguished. Here, we shall avoid this problem, using the symbol $^3Hg*$ to designate $^3P_0$ or $^3P_1$ impartially.

Because spin angular momentum is conserved in bimolecular energy transfer processes, the states of organic molecules populated by collision with $^3Hg*$ are triplet states. The most obvious application of photo-sensitization then resides in the opportunity for selective study of the chemistry of triplet states. A few examples may help to show its place in the scheme of things.

(i) <u>Acetylene.</u> The mercury photosensitized reaction of acetylene leads to polymer (including the cyclic trimer, benzene) and molecular hydrogen. The mode of reaction is

$$^3Hg* + C_2H_2 = Hg + C_2H_2* \qquad (1)$$

followed by,

$$C_2H_2^* = C_2H + H \tag{2}$$

$$C_2H_2^* + C_2H_2 = (C_2H_2)_2; \quad (C_2H_2)_2 + C_2H_2 \rightarrow C_6H_6 \tag{3}$$

with H_2 formation by,

$$H + C_2H_2 = H_2 + C_2H \tag{4}$$

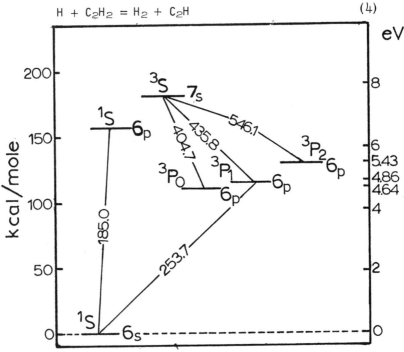

Figure 1.   Energy levels of mercury.

Absorption of the 253.7 nm resonance radiation excites atoms to the $^3P_1$ state. $^3P_0$ is readily formed from $^3P_1$ by collisional deactivation; the direct excitation of ground state mercury into $^3P_0$ is an optically forbidden transition. The relationship of the ^3P states to the familiar 546 (green), 435 (blue) and 405 (violet) nm lines of mercury is illustrated in the upper portion of the Figure.

Formation of linear polymer is probably initiated by attack of C_2H on C_2H_2. It is interesting to note that the triplet state, $C_2H_2^*$, has never been observed in absorption.

(ii) Benzene. The action of $^3Hg^*$ on benzene molecules does not lead to chemical reaction at ordinary temperatures; however, the T_1 ($^3B_{1u}$) state of benzene (85 kcal) can be maintained in sufficient concentration that its optical excitation to a higher triplet state (T_n, $^3E_{2g}$: the serial number n of this state is uncertain) may be observed in absorption[1]. The $T_1 \rightarrow T_n$ triplet-triplet absorption band measured in this way occurs at about 235 nm. At higher temperatures side reactions become appreciable, as shown by the formation of some diphenyl in low quantum yield (Φ = 0.1 at $400^{\circ}C$); evidently some phenyl radicals are produced under these extreme conditions.

The route by which benzene molecules in the T_1 state in the gas phase relax to the ground state has only recently been solved. Direct emission of light is of negligible importance, so that a radiationless process must be involved. The one important step is believed to involve a bi-molecular collision of two molecules in the triplet state, in which both partners are converted into vibrationally-excited ground state molecules ("triplet-triplet annihilation"),

$$2C_6H_6^*(T_1) \quad \rightarrow \quad 2C_6H_6(S_0) \qquad (5)$$

The rate constant measured for this process[1],

$$k_2(\ell \text{ mole}^{-1}\text{sec}^{-1}) = 6 \times 10^{12} \qquad (6)$$

shows that the reaction occurs at essentially every collision. In contrast the lifetime in rare-gas matrices at $4^{\circ}K$, where diffusion is arrested, is about 20 sec.

---

1.   H.E. Hunzinger, IBM J. Res., 15, 10 (1971).

(iii) <u>Olefins</u>. Mercury photosensitization has been used to probe the reactions of ethylenic compounds in the triplet state, especially by observations of the <u>cis-trans</u> isomerization of simple olefins (see Section 3).

2. STATES AND REACTIVITY OF METHYLENE

The methylene radical, CH_2, is formed by photolysis of ketene (240 - 380 nm band), of diazomethane (320 - 470 nm) and similar molecules. The source most frequently used is ketene irradiated with 313 nm radiation from a mercury arc, when the steps involved are,

$$CH_2CO + h\nu = CH_2CO^* \qquad\qquad (7)$$

$$CH_2CO^* = CH_2 + CO \qquad\qquad (8)$$

Because the methylene radicals are formed with excess vibrational energy an inert gas is usually added to "cool" the product to a normal vibrational temperature. Use of excess ketene for this purpose must be avoided because of the further reaction,

$$CH_2 + CH_2CO = C_2H_4 + CO \qquad\qquad (9)$$

In fact, the sequence (7)-(9) in pure ketene leads to a quantum yield of CO formation of about 2.1, the excess over the expected value of 2.0 being due to side reactions.

Spin conservation requires that the CH_2 molecules produced in (8) be in a singlet electronic state. There is no <u>direct</u> evidence that the state so formed is singlet, but it is well established that this state survives only about 500 collisions with inert gas molecules before relaxing to the more stable triplet ground state. The singlet and triplet forms of methylene have different absorption spectra and different physical structure; **singlet methylene has a** strongly bent configuration ($< HCH = 103^{\circ}$), whereas triplet methylene is less strongly bent. The triplet ground state is believed to be about 5 kcal lower in energy than the

138

singlet state, though this value is rather uncertain[2]. In low-temperature matrices methylene shows an e.s.r. spectrum, leaving no doubt that the ground state of the molecule is a triplet state[3].

The rates of reaction of methylene with saturated and unsaturated hydrocarbons, CO, H_2 and other small molecules are very fast, and the relative rates of these reactions have low temperature coefficients. A simple explanation of these observations is that the reaction rates are not much less than the frequency of collisions; otherwise it would be hard to understand how these somewhat dissimilar reactions could have nearly equal activation energies and pre-exponential factors. In the reaction which first led to the direct observation of methylene[4], the results are consistent with the reaction of singlet methylene with diazomethane, Equation (9), being effective at every collision. Whether the vibrationally-excited singlet methylene produced in the step (7) reacts as the 'hot' singlet molecule, as the vibrationally-relaxed singlet, or as the lower-energy triplet species therefore depends on the proportion of inert gas present in the mixture subjected to photolysis.

It is interesting to consider whether a typical reaction of singlet methylene is competitive in rate with the intrinsically-rapid vibrational relaxation. In insertion reactions - where CH_2 replaces a C-H or C-C bond in a saturated hydrocarbon - the answer to this question can be glimpsed in the following way. When ketene vapor is photolyzed in presence of propane there are small but definite differences in the relative rates of insertion into primary and secondary C-H bonds, in the sense that the insertion is about one-sixth more rapid for secondary CH. Diazomethane photolysis, on the other hand, leads to essentially equal

2.   G. Duxbury, J. Mol. Spectrosc., 25, 1 (1968).

3.   E. Wasserman, V.J. Kuck, R.S. Hutton and W.A. Yager, J. Amer. Chem. Soc., 92, 7491 (1970).

4.   G. Herzberg and J. Shoosmith, Nature, 183, 1801 (1959).

rates of insertion. The rationalization for these results is that CH_2 insertion occurs before vibrational relaxation is complete; and that diazomethane photolysis, compared with that of ketene, produces singlet methylene with significantly more vibrational energy, which therefore inserts in an essentially random manner. If relaxation occurred before insertion, no differences are expected. This example is supported by many similar observations.

In the photolysis of ketene mixed with excess ethylene the initial reaction is one of addition, forming cyclopropane. (Vibrational excitation of the product is large, of the order 100 kcal, so that the cyclopropane rearranges to propylene unless it is rapidly "cooled" by inert gas.) Practically every collision is effective in leading to addition, thus the active form is 'hot' singlet methylene. Now, if the ketene photolysis is carried out in presence of cis- or trans-2-butene the addition is stereospecific, that is, cis-2-butene is converted into cis-1,2-dimethylcyclopropane and trans-2-butene into trans-1,2-dimethylcyclopropane.

$$\diagdown\!=\!/ \quad + \ CH_2\text{:} \quad = \quad \triangle \qquad (10)$$

$$\diagdown\!=\!\diagdown \quad + \ CH_2\text{:} \quad = \quad \triangle \qquad (11)$$

Addition reactions of singlet methylene are therefore stereospecific. The complementary experiment is carried out by generating triplet methylene and observing its reaction with olefins. This is achieved by mercury-photosensitized decomposition of ketene which, by the spin-conservation rule, must generate triplet methylene directly. Triplet methylene reacts non-stereospecifically to yield cis- and trans-1,2-dimethylcyclopropane.

$$ (12) $$

The triplet state prepared by this method does not insert into C-H bonds.

3.   CIS-TRANS PHOTOISOMERIZATION

Cis-trans isomerism is of practical importance through its influence on the properties of industrially important materials. Controlled isomerization is thus a means of obtaining materials having desirable characteristics. This section is concerned only with photoisomerization, but it should be realized that isomerization can also be brought about thermally, by acid or base catalysis, or by reaction with odd-electron molecules (e.g. halogen atoms).

Photochemical isomerization can be brought about by direct optical excitation to the π, π^* singlet or triplet states of the olefin, though the practical difficulties are severe. In both of these states the relaxed configuration is 90°-twisted (see Chapter 2) and thus reverts to a mixture of cis- and trans-forms of the ground state. Because the absorption band corresponding to the singlet → triplet excitation is exceedingly weak, isomerization through the triplet state is usually carried out by photosensitization (see below): however, direct photoisomerization of trans-CHD=CHD (1 atm) has been observed after irradiation at 290-340 nm in presence of 130 atm of oxygen, added in order to intensify the absorption. Direct excitation into the singlet excited state is almost equally difficult, because the absorption maximum lies at short wavelengths (165 nm); moreover, the excitation energy is then large (~175 kcal) and unwanted side reactions leading to polymer compete with isomerization.

Mercury photosensitized isomerization, on the other hand, takes place relatively smoothly. Here, the initial process

$$^3Hg^* + \quad C = C \quad (S_0 \text{ state}) = Hg + \quad \overset{.}{C} - \overset{.}{C} \quad (T_1 \text{ state}) \quad (13)$$

must, by the spin conservation rule, produce a triplet state of the olefin. Because $^3Hg^*$ (112 kcal) is much higher in energy than the relaxed T_1 state of olefins (65 kcal), the triplet state must be formed with considerable excess energy; in fact, the mercury photosensitized reaction of cis-CHD=CHD produces some CD_2CH_2 besides trans-CHD:CHD. The mercury sensitized isomerization of cis- and trans-2-butene gives considerable insight into the processes involved. All the features of the reaction are explained by the initial formation of the vibrationally-excited triplet state which either reacts with normal molecules to form polymeric products, or loses the excess vibrational energy by collision. Inert gases (argon, carbon dioxide) are effective in removing vibrational excitation at every collision. In the scheme below, the symbol # denotes excess vibrational energy and M is any molecule contributing to vibrational relaxation.

$$^3Hg^* + \underline{cis}\text{-}CHMe{=}CHMe(S_O \text{ state}) = Hg + CHMeCHMe^\#(T_1 \text{ state}) \tag{14}$$

$$CHMeCHMe^\#(T_1 \text{ state}) \rightarrow \text{polymeric product} \tag{15}$$

$$CHMeCHMe^\#(T_1 \text{ state}) + M = CHMeCHMe(T_1 \text{ state}) + M \tag{16}$$

$$CHMeCHMe(T_1 \text{ state}) = \underline{cis} \text{ or } \underline{trans}\text{-}CHMe{=}CHMe$$
$$(S_O \text{ state}) \tag{17}$$

Deactivation of the T_1 state by (17) yields either isomer with equal probability. The photostationary state actually contains cis- and trans-isomers in the proportion 47:53, for the reason that the cis-isomer is slightly more effective than the trans-isomer as an energy acceptor from $^3Hg^*$.

The corresponding cadmium photosensitized (87 kcal) reaction is slow. Sodium photosensitization (48 kcal) does not produce isomerization. Besides metal atoms, the triplet states of benzene (85 kcal) and sulfur dioxide (73 kcal) are also effective sensitizers for cis-trans isomerization of 2-butene. Triplet biacetyl (56 kcal) is ineffective. All these observations are consistent with the triplet state energy of about 65 kcal deduced from the spectra of simple olefins.

4.   ATMOSPHERIC PHOTOCHEMISTRY

In populated areas, the lower atmosphere is a reaction vessel charged with the by-products of combustion, its chemistry stimulated by sunlight. Details of its reactions vary with geographic location. The type most intensively studied occurs on the west coast of North America where quiet atmospheric conditions, often accompanied by temperature inversion, tend to localize the contaminants within a layer not more than 1500 feet in height, measured from the earth's surface. The typical brownish haze formed in this layer is due mainly to aerosols and particulates formed photochemically by the action of sunlight on the urban atmosphere.

In describing the reactions of this atmosphere it is useful to distinguish the _primary_ pollutants, directly involved in the atmospheric photochemical cycle, from the _secondary_ pollutants which interact with events in the cycle and unbalance it to the extent that steady-state conditions do not prevail. These primary pollutants are small molecules, namely, oxides of nitrogen and ozone. The nitrogen oxides are discharged, chiefly as NO but with some admixture of NO_2 and HNO_3 vapor, from automobile exhausts and from all furnaces or other stationary installations burning fossil fuel. Because the high temperature of the gasoline engine favors the formation of nitrogen oxides as secondary products of combustion, the automobile accounts for about 60% of all nitrogen oxide emission in the Los Angeles area. The fact that the automobile is a mobile generator then helps to diffuse pollution by these oxides over a broad territory. In the photochemical cycle the following reactions occur:

$$NO_2 + h\nu(300 - 400 \text{ nm}) = NO + O(^3P) \tag{18}$$

$$O + O_2 + M = O_3 + M \tag{19}$$

$$O_3 + NO = O_2 + NO_2 \tag{20}$$

Thus the effect of sunlight on the NO_2 - O_2 organic system is to establish an equilibrium,

$$NO_2 + O_2 \overset{h\nu}{\rightleftarrows} NO + O_3 \tag{21}$$

providing a regenerative supply of powerful oxidizing agents for reaction with the secondary pollutants. The characteristic property of this atmosphere is its strong oxidizing action.

Although the steps (18) - (20) are individually simple, there are certain aspects which deserve more detailed consideration. Sunlight striking the lower atmosphere is deficient in wavelengths <300 nm because this radiation is absorbed by the protective ozone layer at higher levels. Photolysis of nitrogen dioxide does not occur if the wavelength absorbed is >398 nm (72 kcal), essentially equal to the dissociation energy of nitrogen dioxide into NO + O in their electronic ground states. The active photochemical wavelengths are therefore in the ultraviolet region, approximately 300 - 400 nm. Rate constants for some of the processes involved in atmospheric photochemistry are summarized in Table I[5], where the first two rows contain data for reactions involved in the primary cycle.

The reactions listed in Table I are all to be considered as "fast" reactions, though not so fast that they occur at every collision. The collision rate, which cannot be specified exactly since it depends on the effective cross-section (diameter) of the colliding partners, is approximately 10^{11} - 10^{12} ℓ mole^{-1} sec^{-1} at ordinary temperatures. For most of the processes listed, therefore, reaction occurs once in about 10^2 - 10^4 collisions. The first, third and final rows refer to three-body processes whose (termolecular) rate constant has the units ℓ^2 mole^{-2} sec^{-1}. In these equations M stands for a molecule of O_2 or N_2 whose function is to stabilize the product by accepting some part of the energy liberated by the reaction. The concentrations of O_2 and N_2 in the atmosphere together amount to about 0.04 mole liter^{-1}: thus the rate constants for the three-body reactions can be cited as pseudobiomolecular rate constants (ℓ mole^{-1} sec^{-1}) by multiplying the termolecular constant by 0.04. This rough and ready procedure, which takes no account of possibly different efficiencies on the part of O_2 and N_2 as stabilizers, allows for a direct comparison between ter- and bimolecular reactions.

---

5.   R.D. Cadle and E.R. Allen, Science, 167, 243 (1970).

TABLE 1

Reaction	k_2, ℓ mole^{-1} sec^{-1}
$O + O_2 + M = O_3 + M$	10^7 (pseudobimolecular)
$O_3 + NO = O_2 + NO_2$	3×10^7
$O + O + M = O_2 + M$	4×10^7 (pseudobimolecular)
$O + RCHO = HO + CH_3CO$	3×10^8
$O_3 + olefin = products$	$2 - 20 \times 10^3$
$O + olefin = products$	$1 - 5 \times 10^9$
$R + O_2 = RO_2$	$\sim 10^7$
$SO_2 + O + M = SO_3 + M$	2×10^8 (pseudobimolecular)

It should be stressed that a rate constant represents a reaction rate normalized to unit concentration of reactants: it is the rate, in moles liters^{-1} sec^{-1}, that would be realized at a fixed reactant concentration of 1 mole liter^{-1}. Results in the Table show that the rate constants for the processes,

$$O + O_2 + M = O_3 + M \qquad (19)$$

and

$$O + O + M = O_2 + M \qquad (22)$$

are slightly in favor of the latter reaction. But the
rates of the two processes are overwhelmingly in favor of
the former, because the concentration of O_2 is very much in
excess over that of O atoms.

The secondary pollutants are varied; not all of them
participate directly in forming the photochemical smog. A
typical profile for pollutants of all types appears in
Table II where the concentration is stated in parts by
volume per hundred million parts of air (pphm: 1 pphm represents a concentration of 4×10^{-10} mole liter^{-1})[5]. The
automobile is the chief source of hydrocarbons and carbon
monoxide, whereas stationary burners are responsible for
most of the sulfur dioxide contamination. Following the
1966 regulations governing automobile emission, the concentration of CO in urban atmospheres has tended to level
off instead of showing the increase which might otherwise
have been expected. Unfortunately, this advance seems to
have been made at the expense of an increase in the emission
of nitrogen oxides and aldehydes, both of which contribute
actively to smog formation.

Atmospheric pollution in the east and mid-west, as
well as in Europe, involves much higher concentrations of
sulfur dioxide than those which occur in the Los Angeles
basin. In these areas the sulfur dioxide confers reducing
properties instead of the oxidizing property characteristic
of west coast conditions. Its properties as a reducing
agent lead to the formation of trioxide (as sulfuric acid)
which causes widespread economic damage by corrosive action
on structural stone and metals.

The reactions (18) - (20) are non-branching. If no
other processes occurred, the result would be a steady-
state concentration of ozone considerably smaller than that
which actually occurs in the polluted atmosphere. The
branching mechanism is not understood in detail but is
believed to occur in cycles initiated by the action of
ozone on the chief secondary pollutants, hydrocarbons and

TABLE II

Concentration of pollutants in photochemical smog
under typical conditions

	Concentration (pphm)
Nitrogen oxides	20
Carbon monoxide	4000
Carbon dioxide	40000
Ozone	50
Methane	250
Higher alkanes	25
Ethylene	50
Higher alkenes	25
Aromatic hydrocarbons	10
Aldehydes	60
Sulfur dioxide	20

aldehydes. Both derive from fuels, by incomplete combustion or by escaping combustion altogether, and aldehydes are also produced in the atmosphere by the attack of ozone on unsaturated hydrocarbons. In one mode of reaction, aldehydes react with atomic oxygen to produce acyl radicals,

$$O + RCHO = RCO + OH \qquad (23)$$

The acyl radicals react with O_2 to form peroxyacyl radicals which are capable of oxidizing nitric oxide, probably by a regenerative chain reaction. The termination of these chains results in formation of peroxyacylnitrate (PAN),

$$RCO.O_2 + NO_2 = RCO.O.ONO_2 \qquad (24)$$

Other chain reactions are thought to be initiated by attack of atomic oxygen or ozone on olefins or reactive saturated hydrocarbons, forming organic free radicals which combine rapidly with molecular oxygen to produce peroxy free radicals,

$$R + O_2 = RO_2 \qquad (25)$$

RO_2 radicals are capable of oxidizing nitric oxide to nitrogen dioxide and may be responsible for a significant part of the nitrogen dioxide present in the polluted atmosphere. A similar reaction probably accounts for a considerable proportion of the ozone produced,

$$RO_2 + O_2 = RO + O_3 \qquad (26)$$

Aldehydes and other intermediates absorb sunlight in the ultraviolet region and may contribute independently to the atmospheric reaction mixture.

The damaging effects of atmospheric pollution, though not directly a photochemical problem, are important enough to be mentioned briefly. Sulfur dioxide, ozone and PAN are toxic to many forms of plant life, including important agricultural crops, so that in areas of photochemical pollution the latter two cause widespread damage. Nitrogen dioxide appears to retard growth, especially in young plants, without producing visible signs of injury. The cumulated effects on human life may also be considerable. A prolonged fog in London in 1952 is considered to have caused several thousand deaths, mainly among people already suffering from bronchial disease. It is thought that neither sulfur dioxide nor the particulates (mainly, carbon) were separately responsible, rather that their simultaneous presence led to synergistic effects in which the combination of finely-divided carbon plus sulfur dioxide was more deadly than the sum of parts. The health implications of air pollution are much more complex than its chemistry.

CHAPTER 7

MECHANISMS OF ORGANIC PHOTOCHEMICAL REACTIONS

Introduction

One of the most striking changes which has occurred in organic chemistry in the last decade is the recognition that molecules in electronic excited states may undergo useful and interesting reactions. It is now clear that, since the reactivity of the electronic excited state is different from that of the ground state, unusual compounds, difficult to prepare from ground state reagents, can be made photochemically. The main drawback to the immediate extensive use of photochemistry for the synthesis of complex molecules is the dearth of rigorous generalizations with regards the scope and limitations of the potentially useful reactions. Questions pertaining to what functional groups will interfere with a given photosynthetic transformation; what protecting groups can be used to prevent this interference; and what may be the effects of substituents and solvents, remain largely unanswered. Such generalizations have traditionally come from a detailed understanding of the reaction, couched in mechanistic terms.

The task of understanding a photochemical reaction is more demanding than that for a ground state reaction for several reasons. Most molecules have more than one excited state available (and many have several). Each type may be of different multiplicity, the singlet and triplet being by far the most important. The first problem, therefore, is to define which excited state is the reactive "reagent". It is also clear that the same type of product can arise from different types of excited states, and also from one state, but by more than one mechanism.

An additional feature, not parallelled in ground state chemistry is that knowledge of the competing deactivation processes is also required. These latter are the photophysical phenomena such as intersystem crossing, internal conversion and energy transfer discussed in previous chapters. Frequently, these processes are too fast to allow a bimolecular reaction to occur; that is, even if the reactive state is obtained, its lifetime is too short for a reaction requiring collision.

In addition to these problems, inherent in a photochemical process, it is frequently difficult to define at what stage along the reaction coordinate the photochemistry is over and the reaction has become one involving ground state intermediates. This difficulty may be enhanced if these intermediates, although in the ground electronic state, are without analogy in thermal reactions. Thus, much of the undetermined in photochemistry may, in fact, lie within the realm of unusual ground state intermediates.

This chapter will summarize five general and typical reactions: hydrogen abstraction, bond cleavage, cyclo-addition, valence isomerization and rearrangements. The examples have been chosen to illustrate the scope of each reaction and to give some idea how generalizations can be derived. There are several text books, listed at the end of this chapter, which compile a more complete list of reactions and a new volume which gives a detailed discussion of experimental procedures.

Hydrogen abstraction

One of the oldest and most extensively studied photochemical reactions is that of hydrogen atom abstraction. The transfer occurs from a ground state molecule to a molecule in an excited state with the resultant formation of two radicals. For example, upon irradiation, carbonyl compounds can abstract a hydrogen atom from several classes of compounds. These include alkanes, alkenes, alcohols, and ethers, and in all cases, the resulting radicals then react further, usually by disproportionation or coupling.

Illustrative of this process is the formation of benz-pinacol by irradiation of benzophenone in isopropyl alcohol. This reaction was first described (amongst many 'firsts') by Ciamician and Silber at the turn of the century. Under certain conditions the reaction pathway can be described by the series of steps shown, briefly, in Scheme (1).

Scheme   (1)

The Reduction of Benzophenone upon Irradiation

in Isopropyl Alcohol

Absorption of light

$$C_6H_5COC_6H_5 \xrightarrow{h\nu} C_6H_5COC_6H_5$$
$$(n, \pi^* \text{ or } \pi, \pi^* \text{ singlet})$$

Intersystem crossing

$$C_6H_5COC_6H_5 \rightarrow C_6H_5COC_6H_5$$
$$(n, \pi^* \text{ or } \pi, \pi^* \quad (n, \pi^* \text{ triplet})$$
$$\text{singlet})$$

Hydrogen abstraction

$$C_6H_5COC_6H_5 + CH_3C(OH)HCH_3 \rightarrow$$
$$(n, \pi^* \text{ triplet})$$

$$C_6H_5\overset{\bullet}{C}(OH)C_6H_5 + CH_3\overset{\bullet}{C}(OH)CH_3$$

Hydrogen transfer

$$CH_3\overset{\bullet}{C}(OH)CH_3 + C_6H_5COC_6H_5 \rightarrow CH_3COCH_3$$

$$+ C_6H_5\overset{\bullet}{C}(OH)C_6H_5$$

Coupling

$$2C_6H_5\overset{\bullet}{C}(OH)C_6H_5 \rightarrow (C_6H_5)_2\overset{\overset{\displaystyle OH}{|}}{C} - \overset{\overset{\displaystyle OH}{|}}{C}(C_6H_5)_2$$

152

On the basis of the kinetics implied by these equations, it may be shown that the intersystem crossing efficiency is near unity, and rate constants can be found for radiationless decay of the triplet and for its reaction with isopropanol. This is a typical chemical approach to the characterization of the reactive species.

One of the most useful generalizations that has emerged from the study of the reduction of carbonyl compounds by this mechanism is the following: carbonyl compounds which have a lowest triplet with n, π^* character will be reactive whereas π, π^* triplet states are relatively unreactive. Thus, acetone, benzaldehyde, acetophenone, benzophenone, and many analogs are reduced to the pinacol under these conditions, but this reaction becomes inefficient with any of the naphthyl carbonyl compounds where the lowest triplet has π, π^* character. The same is true of amino-substituted ketones, such as $4,4'$-bis(dimethylamino)-benzophenone where the lowest triplet has π, π^* (charge-transfer) character.

Correlations between photochemical reactivity and emission spectra can be quite useful inasmuch as the latter may serve to define the relevant state. It must be kept in mind, however, that the conditions of these two types of experimentation are very different. The photochemical reaction is usually carried out at room temperature in fluid solution while the emission spectra, at least until recently, have been measured at low temperature in rigid media. State diagrams illustrating reactive and unreactive carbonyl compounds are shown in the Table following.

153

Partial State Diagrams for Some Carbonyl Compounds

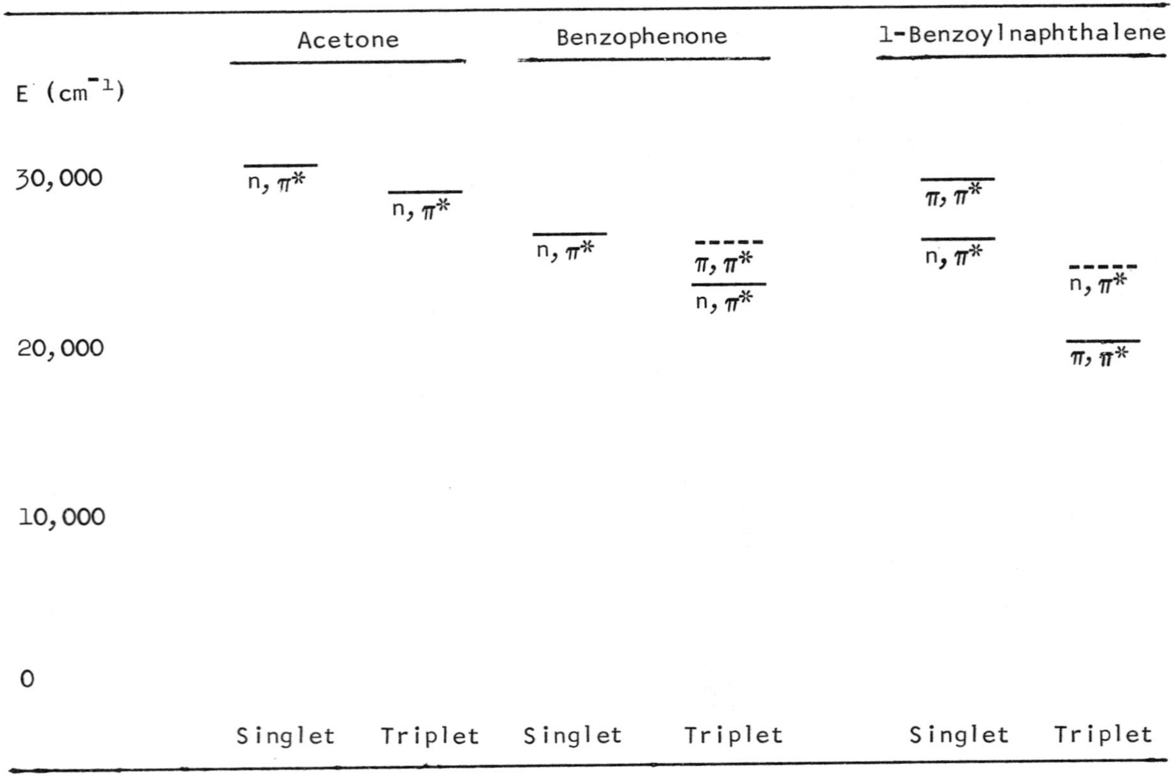
Acetone
Benzophenone
1-Benzoylnaphthalene
E (cm⁻¹)
30,000
20,000
10,000
0
n, π*
n, π*
n, π*
π, π*
n, π*
π, π*
n, π*
n, π*
π, π*
Singlet
Triplet
Singlet
Triplet
Singlet
Triplet

The reactivity of the n, π^* state toward hydrogen abstraction is, in part, understandable in view of the structure of this state. Excitation of one of the nonbonding electrons on the oxygen atom into a π^* orbital leaves an unpaired electron on oxygen. In that sense the excited state then has alkoxy radical character.

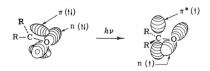

This is of course true for both the n, π^* singlet and triplet so both should be reactive in the hydrogen abstraction reaction. Usually the singlet state is too short-lived to be an important reactant at least in bimolecular reactions, but if intersystem crossing is slow, then reaction from this state may well be observed.

Intersystem crossing is faster between states of different symmetry; for example, when the lowest singlet is n, π^* which can cross to a lower π, π^* triplet and then on to an even lower n, π^* triplet (if one exists). This is frequently the case with aromatic carbonyl compounds. Aliphatic carbonyl compounds with a lowest n, π^* singlet can cross only to an n, π^* triplet; thus, the singlet is longer lived. Perhaps for this reason the n, π^* singlet state of aliphatic carbonyl compounds appears to undergo bimolecular reactions while only the triplet state is involved with aromatic carbonyl compounds.

There are several fairly general reactions of carbonyl compounds that involve hydrogen abstraction as the first step. Amongst the most studied is the reaction which is intramolecular and in which the hydrogen transferred is on the carbon γ to the carbonyl group.

155

This is the Norrish type II process, an intramolecular hydrogen abstraction process that competes with intermolecular reaction.

Ortho alkyl benzophenones, possessing such γ hydrogen atoms, can be used as ultraviolet stabilizers in plastics. The explanation for their protective property is that hydrogen transfer is equivalent to photoenolization. This is followed by ketonization, a non photochemical reaction, back to starting material. The overall sequence represents a way of degrading otherwise harmful light energy into heat with no overall chemical change. This photoenol can be

156

trapped with good dienophiles.

Allylic hydrogens are easily abstracted and the resulting radicals can couple to yield the alcohol.

$$CH_3-\overset{O}{\underset{\|}{C}}-CH_3 \ + \ \bigcirc \ \xrightarrow{h\nu} \ [CH_3\overset{\cdot}{C}(OH)CH_3] \ + \ \left[\bigcirc\right] \longrightarrow \overset{\overset{OH}{|}}{\underset{}{H_3C-\overset{|}{C}-CH_3}}\!\!\bigcirc$$

α-Diketones and α-ketoacids and esters can also abstract hydrogen.

This sequence illustrates the importance of ground state considerations in an apparently purely photochemical reaction. The formation of the ketyl radical is photochemical, but the products differ considerably depending on the temperature. This is entirely a question of the thermal stability of the ketyl radical, and in no way reflects the presence of an activation energy in the photochemical pathway.

Carbonyl compounds are not unique in forming reactive n, π^* states; nitro-aromatic compounds are also reduced upon irradiation in hydrogen-donating solvents. This reaction is apparently not useful for the preparation of the reduced nitro compound because of the variety of products obtained; however, some use may develop for this reaction as an oxidizing step for the hydrogen donor.

$$C_6H_5NO_2 \; + \; \bigcirc \; \xrightarrow{\; h\nu \;} \; \overset{OH}{\bigcirc} \; + \; \overset{O}{\bigcirc} \; + \; \text{(mixture of reduction products from } C_6H_5NO_2 \text{)}$$

Imines are reduced to ethylenediamines upon irradiation in ethanol. This reaction was initially believed, in part by analogy with carbonyl compounds, to involve the n, π^* state of the imine. Analogy is an even more dangerous tool in photochemistry than in ground state chemistry and it has recently been shown that the imine excited states are not involved. Instead the reaction occurs by a hydrogen transfer mechanism in which the n, π^* triplet state of the carbonyl compound (either added as a sensitizer or formed by hydrolysis of the imine by adventitious water) initiates the reaction.

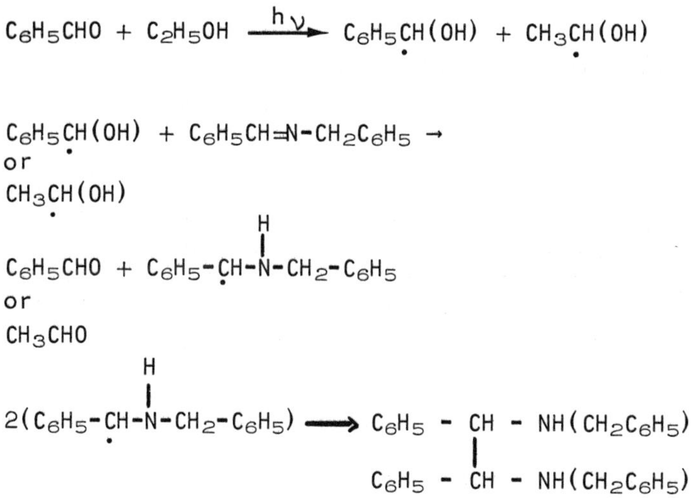

$$C_6H_5CHO + C_2H_5OH \xrightarrow{h\nu} C_6H_5\overset{\cdot}{C}H(OH) + CH_3\overset{\cdot}{C}H(OH)$$

$$C_6H_5\overset{\cdot}{C}H(OH) + C_6H_5CH\!\!=\!\!N\!-\!CH_2C_6H_5 \rightarrow$$
or
$$CH_3\overset{\cdot}{C}H(OH)$$

$$C_6H_5CHO + C_6H_5\!-\!\overset{\overset{\displaystyle H}{|}}{\underset{\cdot}{C}H}\!-\!N\!-\!CH_2\!-\!C_6H_5$$
or
$$CH_3CHO$$

$$2(C_6H_5\!-\!\overset{\overset{\displaystyle H}{|}}{\underset{\cdot}{C}H}\!-\!N\!-\!CH_2\!-\!C_6H_5) \longrightarrow \begin{array}{c} C_6H_5 - \overset{|}{C}H - NH(CH_2C_6H_5) \\ | \\ C_6H_5 - CH - NH(CH_2C_6H_5) \end{array}$$

The discussion of hydrogen abstraction thus far has involved species possessing the n,π^* excited state, either singlet or triplet. It is now clear that other types of excited states may also abstract hydrogen. The generalizations governing these reactions have not yet been developed and in some cases a different mechanism (i.e. electron transfer followed by proton transfer) may be involved. For instance, the following transformation has been observed:

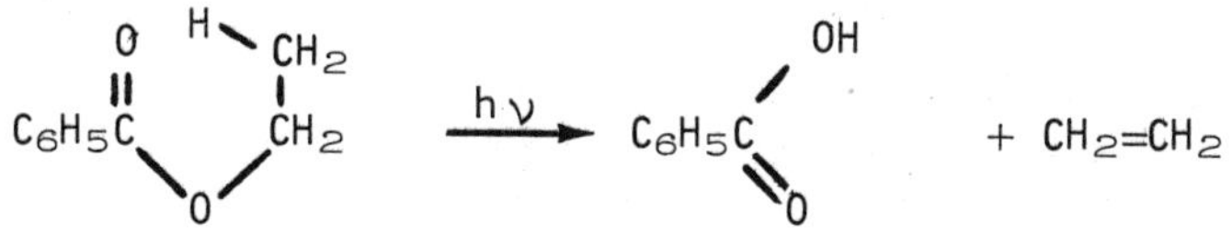

The lowest singlet and triplet state of benzoate esters have however, π,π^* character, yet, formally at least, the Norrish type II cleavage occurs.

It has been reported that 1,1-diphenylethylene is reduced upon irradiation in isopropyl alcohol. In this case a π,π^* state is necessarily involved.

$$(C_6H_5)_2C{=}CH_2 \ + \ CH_3{-}CH(OH){-}CH_3 \ \xrightarrow{h\nu} \ \begin{array}{c} (C_6H_5)_2{-}C{-}CH_3 \\ | \\ (C_6H_5)_2{-}C{-}CH_3 \end{array}$$

A more complex system involves an amino-ketone, 4,4'-dimethylaminobenzophenone. Even though this substance is not reduced upon irradiation in isopropyl alcohol, reaction does occur in the presence of benzophenone. The reaction apparently proceeds as follows:

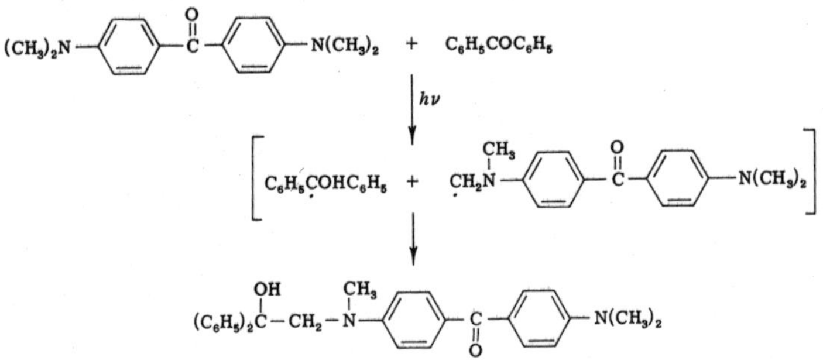

The presence of the intermediate ($\overset{\bullet}{C}H_2{-}\overset{\overset{\textstyle CH_3}{|}}{N}{-}$) radical was demonstrated by the exchange of the N-methyl protons with deuterium upon irradiation in the presence of deutero-ethylmercaptan.

The mechanism for this reaction is believed to involve a complex formed between the excited state of one ketone molecule and another ketone molecule in the ground state, an exciplex. In fact, exciplexes are now believed to be involved as the first formed intermediate in several photochemical reactions, and are undoubtedly the demons behind the demise of several "generalizations" concerning excited state reactivity. More will be said about exciplexes later.

160

Bond cleavage reactions

When an organic molecule absorbs light in the ultraviolet region it acquires enough energy to break all but
the strongest bonds: light of wavelength 286 nm is equivalent to 100 kcal/mole. It is thus, perhaps, surprising
that the excited molecule does not dissociate into radicals
in an indiscriminate fashion. In fact, most molecules do
undergo some fragmentation when the irradiating light is of
sufficiently short wavelength, but more energy is required
than the minimum bond dissociation energy. This is because
part of the excitation energy is usually lost in the form
of increased vibrational energy (i.e. heat) to the surround-
ings. This accounts for the fact that bond homolysis is
much more common and less selective in the vapor phase
where, in contrast to excitation in condensed phases, the
energy required cannot be so easily dissipated rapidly
through collision with other molecules.

Bond homolysis in solution is usually specific with
the weakest bond close to the absorbing chromophore cleav-
ing selectively to give the most stable radical. The exact
mechanism of the homolysis step is known for only the
simplest molecules. With complex molecules, cleavage is
known to occur following initial excitation into several
different types of chromophores (e.g. n, π^*, π, π^*, σ, σ^*) and
from both the singlet and triplet states. The ultimate
product obtained depends upon the subsequent behaviour of
the ground state radicals (or diradical) that are formed,
and several synthetically useful reactions have here a
common origin. A selection of such useful reactions follows.

The formation of nitrenes by photolysis of azides is a
generally utilized method of obtaining these reactive intermediates.

$$\text{(benzene)} + \text{N}_3-\text{CO}_2\text{C}_2\text{H}_5 \xrightarrow{h\nu} \text{(azepine)}\text{N}-\text{CO}_2\text{C}_2\text{H}_5$$

The formation of aryl radicals may be induced by the bond homolysis of aryl iodides. This may be a useful source of aryl radicals for homolytic arylation, as shown in the following example:

$$\text{(4-iodophenol)} \xrightarrow[\text{C}_6\text{H}_6]{h\nu} \text{(4-phenylphenol)}-\text{OH}$$

Irradiation of o-diiodobenzene leads to benzyne formation.

The Norrish Type 1 process is the cleavage of a carbon bond α to a carbonyl group. Ordinarily that bond cleaves which leads to the formation of the most stable radicals. The ultimate product, however, depends upon the reactivity of these radicals. In certain cases the intermediates may be trapped, as shown in the following example involving cyclobutanone and butadiene.

163

In other examples intramolecular hydrogen transfer may occur rapidly enough to compete with loss of carbon monoxide which otherwise is the most usual outcome. One frequent mode of hydrogen transfer, illustrated below, leads to ketene formation. The latter, in the presence of alcohols gives the corresponding esters.

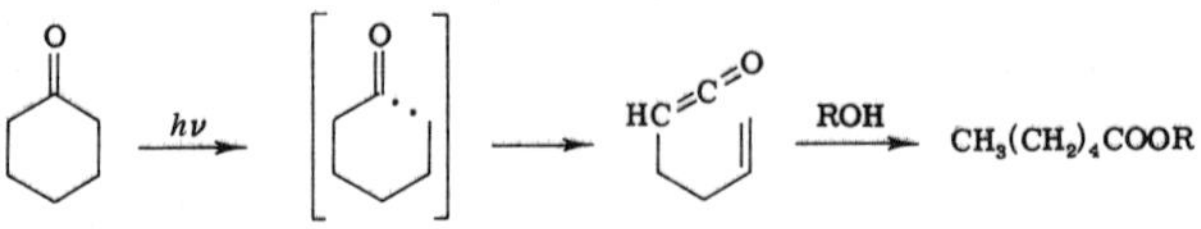

Irradiation of azo compounds and peroxides can lead to free-radical formation. For instance, azobisisobutyronitrile (AIBN) is a common radical chain initiator.

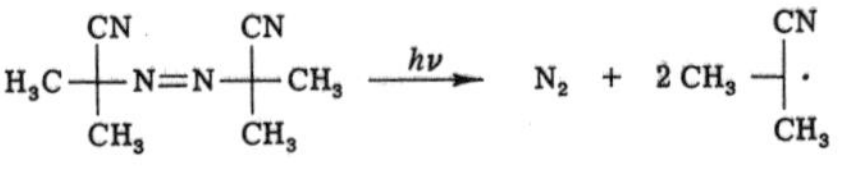

The photolysis of nitrogen and oxygen halides and pseudohalides also generates radicals. The Barton reaction, the irradiation of an alkyl nitrite ester to give an oxime alcohol, is an example which found spectacular use in the partial synthesis of aldosterone.

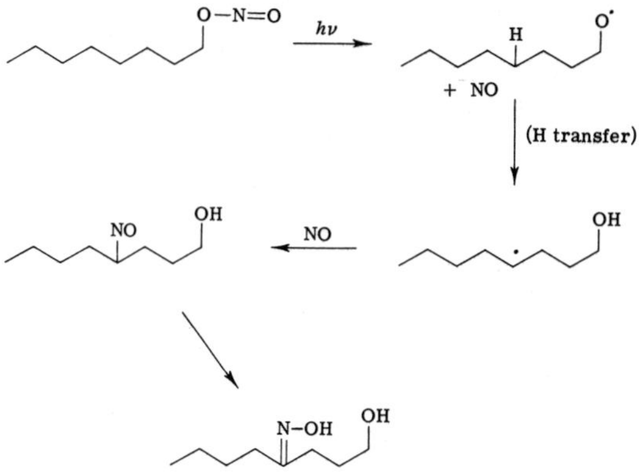

O—N=O $h\nu$ →   H   O·

+ NO

(H transfer)

NO   OH   ← NO ←   OH

N—OH   OH

A similar process occurs on the irradiation of hypochlorites.

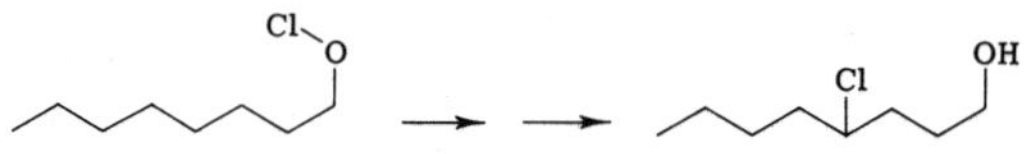

Cl—O   → →   Cl   OH

Cycloadditions

One of the most useful and extensively studied photochemical reactions is the photocycloaddition of one unsaturated system to another. These reactions are particularly important in synthesis since four-membered rings, difficult to prepare by classical ground state methods, are quite commonly formed. Examples of six- and eight-membered ring formations are also known. The procedure may have the added synthetic value that carbon-carbon bond formation may occur at an unactivated site.

Several different chromophores can participate in these reactions and more than one mechanism can be involved with the same chromophore. It is thus apparent that **several** sets of generalizations (and many provisos) will be necessary to predict reactivity.

The two reactions which have received the most attention are the photocycloaddition of carbonyl compounds to olefins to give oxetanes and the photocycloaddition of one olefin or conjugated olefin to another. Several other types of cycloadditions are also known and some examples will be illustrated below.

Photocycloaddition of carbonyl compounds to olefins

In connection with the photoreduction of ketones, it was pointed out that the carbonyl n,π^* state has alkoxy radical character. It is thus not surprising, since alkoxy radicals are known to add to olefins, that the cycloaddition of carbonyl compounds to olefins can involve the n,π^* state. The reaction was first assumed to proceed by addition of the electron-deficient oxygen to the olefin to give a 1,4-biradical:

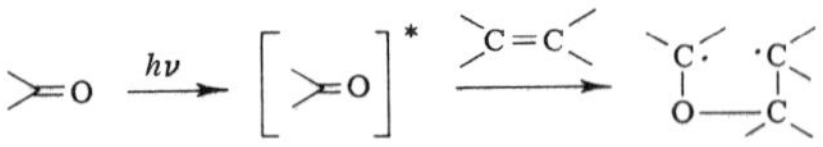

However the reaction appears to proceed with a rate constant ($10^8 M^{-1} sec^{-1}$) much faster than that expected for ordinary alkoxyl radical attack. For this and other reasons it seems probable that in certain cases another intermediate - an exciplex or excited state complex - may precede the biradical as intermediate.

Both the n,π^* singlet and triplet can take part in this reaction. With aromatic ketones, the rapid rate of intersystem crossing precludes reaction of the singlet, but several examples are known where the n,π^* singlet of aliphatic ketones and aldehydes is involved.

Benzophenone, on excitation (n,π^* triplet), will add to isobutylene:

$$C_6H_5CC_6H_5 \ (O) \quad + \quad CH_3 \underset{CH_3}{\overset{CH_2}{\diagup}} \quad \xrightarrow{h\nu} \quad (C_6H_5)_2 \ \square \ (CH_3)_2$$

The predominant product is that expected from the most stable diradical intermediate, and this appears to be a general observation.

The same benzophenone n,π^* triplet adds to norbornene to give the exo-adduct.

$$C_6H_5CC_6H_5 \quad + \quad \xrightarrow{h\nu} \quad (C_6H_5)_2$$

If the olefin involved in the reaction has easily abstractable hydrogen atoms, reduction of the benzophenone competes with cycloaddition. This seems to be the case when cyclohexene is employed as substrate.

In the photocycloaddition of acetone to cyclooctene, the n,π^* triplet, as in the case of benzophenone and re-

167

lated substances, is the major reactive state. The cycloaddition is not stereospecific since the 1,4-diradical is,
here also, an intermediate.

CH_3COCH_3 + (cyclooctene) $\xrightarrow[n,\pi^* \text{ triplet}]{h\nu}$ (bicyclic oxetane) $(CH_3)_2$ + (bicyclic oxetane) $(CH_3)_2$

CH_3COCH_3 + (cyclooctadiene) $\xrightarrow[n,\pi^* \text{ singlet}]{h\nu}$ (bicyclic oxetane) $(CH_3)_2$ + (bicyclic oxetane) $(CH_3)_2$

On the other hand, when 1,3-cyclooctadiene is used the
addition involves the n,π^* singlet of acetone. The n,π^*
triplet energy of acetone is greater than the triplet energy
of the diene so that the latter acts as a quencher for the
acetone triplet. The singlet is not quenched and is available for chemical reaction.

$C_6H_5\overset{O}{\overset{\|}{C}}CH_3$ + (cyclooctadiene) $\xrightarrow{h\nu}$ Cyclooctadiene dimers
(no oxetane is formed)

When the corresponding reaction is attempted using acetophenone and 1,3-cyclooctadiene the 1,3-cyclooctadiene dimers
are the major product and no oxetane is formed. This is
because here the rapid intersystem crossing rate of the
aromatic ketones precludes bimolecular reaction of the n,π^*

singlet. The n, π^* triplet of acetophenone, like that of acetone, is quenched by the diene giving the π, π^* triplet of the diene. This then reacts with the ground state diene to give the dimers.

While most cases of cycloaddition of the carbonyl group involve the n, π^* state, there seems little reason to exclude <u>a priori</u> the π, π^* state as a possible reactant, particularly since the π, π^* state of olefins is known to be responsible for cyclobutane formation. It is therefore interesting that it has recently been shown that irradiation of some substituted methyl benzoates in the presence of olefins gives alkoxy oxetanes. This reaction may well involve a π, π^* state of the ester since both the lowest singlet and triplet are π, π^* states. More work will be required in order to define the scope of the reaction via these diverse mechanisms.

Much the same situation pertains with regard to the photocycloaddition of one olefin to another. Examples of reactions involving both the π, π^* singlet and triplet are known and exciplexes are frequently believed to be involved.

The π, π^* singlet of trans-stilbene is the reactive state in the addition to tetramethylethylene. Kinetic analysis indicates that an exciplex is the first formed intermediate, and the reaction has been shown to be stereospecific. The π, π^* triplet, obtained by triplet-triplet transfer is not reactive presumably because cis-trans isomerism occurs too rapidly and provides an energy wasting mechanism which competes with photocycloaddition.

$$\begin{array}{c} C(CH_3)_2 \\ \| \\ C(CH_3)_2 \end{array} \xrightarrow{h\nu} \begin{array}{c} (CH_3)_2 \\ (CH_3)_2 \end{array}\square\begin{array}{c} (CH_3)_2 \\ (CH_3)_2 \end{array}$$

The photodimerization of simple olefins (tetramethyl-ethylene and 2-butene) also occurs only upon direct irradiation.

88% (sens)   12% (sens)
100% (direct)

With cyclic olefins, however, reaction occurs from both the π, π^* singlet and the π, π^* triplet. The ratio of the dimers from norbornene is different for each state. The π, π^* triplet of cyclic olefins with ring size less than seven will form dimers presumably because the cis-trans deactivation pathway is prohibited; larger cyclic and acyclic olefins generally undergo cis-trans isomerization, but there are exceptions which indicate the need for more study. For example, the photocycloaddition of diphenylethylene to isobutylene is reported   occur from the triplet state of diphenylethylene.

$$(C_6H_5)_2C{=}CH_2 \quad + \quad (CH_3)_2C{=}CH_2 \xrightarrow{\ h\nu\ } (C_6H_5)_2 \ \square \ (CH_3)_2$$

The following examples will further indicate the scope of the photocycloaddition reaction.

172

Valence isomerization

The interconversion of valence isomers is another general and useful photochemical reaction. These reactions usually involve bond making and breaking in a concerted fashion and thus are controlled by orbital symmetry considerations. We introduced this concept briefly in Chapter 2; for a more detailed discussion, see one of the excellent recent reviews on this subject referred at the end of the Chapter.

Some idea of the scope of this reaction can be gained from the following examples. These reactions are often photochemically reversible. One isomer generally predominates at photochemical equilibrium because of the large differences in absorption spectrum, but if the product is optically transparent, then the reaction will not be photochemically reversible and will proceed to completion.

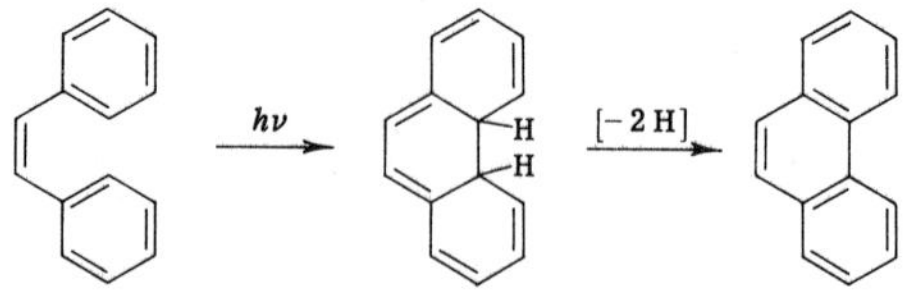
hν
H
H
[−2 H]

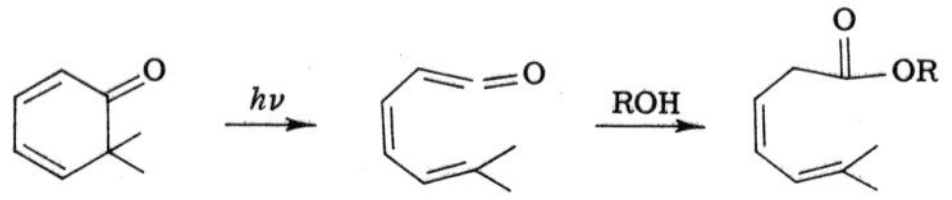
O
hν
O
ROH
O
OR

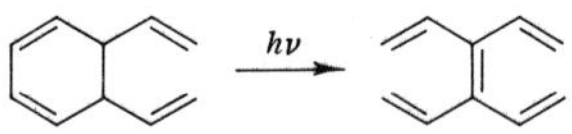
hν

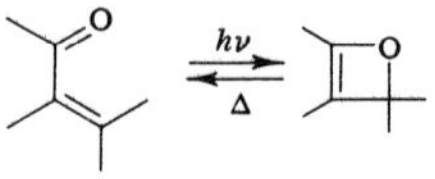
O
hν
Δ
O

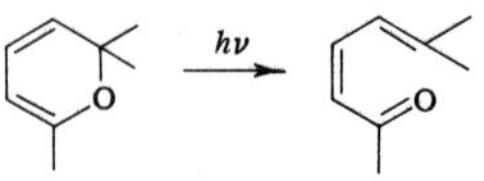
O
hν
O

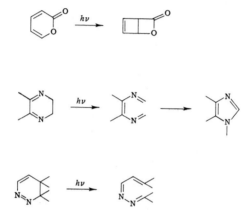

The formal analogy of these reactions is obvious;
however, there the known similarity ends. In fact, mechanistic understanding in most cases is at a very early stage
of development. Excitation of several different chromophores can bring about reaction and more than one mechanism
is undoubtedly involved.

Reactions 1 - 5 are examples of the diene to cyclobutene or bicyclobutane isomerization. Cyclohexadiene and
hexatriene are also interconverted (reactions 4, 7 - 9)
in an equivalent vinylogous reaction. Reactions 10-14 give
some examples of heterocyclic systems that behave similarly.
For most of the cases studied, the isomerization can occur
from the π, π^* singlet state; other states may or may not
lead to reaction. Excitation of 3,4-dimethyl-3-pentene-2-
one (reaction 10) to the $n \rightarrow \pi^*$ state is not sufficient to
bring about the reaction; excitation of the $\pi \rightarrow \pi^*$ transition
is required.

Sensitized excitation of cyclohexadiene or cyclobutene
to the π, π^* triplet gives, not the valence isomerization,
but the dimers expected of the photocycloaddition reaction.

Rearrangement

There are many different types of photochemical rearrangements, and many of these involve complex mechanisms
with several intermediates. The first step may, in fact, be
one of the photochemical reactions already mentioned, where
the first formed product subsequently rearranged non-photo-
chemically. Only a very few of the most general types can
be discussed here.

One such transformation is that now termed the di-π-
methane rearrangement, and three examples are illustrated.
These reactions proceed in high yield and are thus preparatively useful. They derive from the singlet state.

176

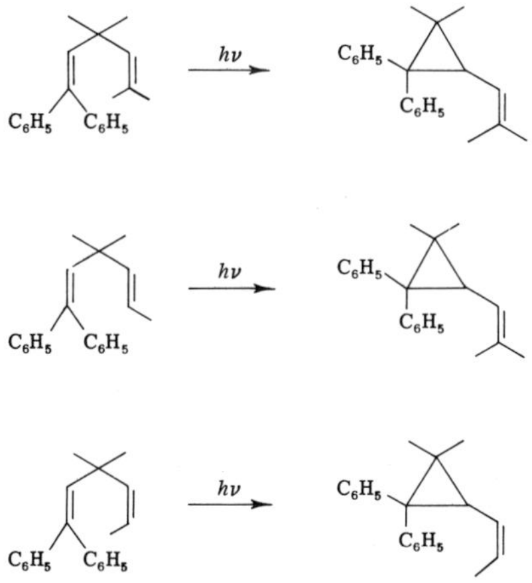

When the starting dienes are excited indirectly to the triplet state by triplet-triplet energy transfer no such rearrangement occurs; instead cis-trans isomerization is observed. A mechanistic interpretation of these results is illustrated below. While possible intermediates are shown the reaction may well proceed by a concerted pathway.

A fascinating corollary to this series is the formally analogous oxa-di-π-methane rearrangement.

In this case the reaction proceeds in high yield but now only from the triplet state. Direct irradiation leads to completely different products. The overall mechanism may in fact be very similar, the main difference being in the rates of competing reaction and deactivation pathways.

Cyclic conjugated enones and cross-conjugated dienones undergo a variety of interesting complex transformations upon irradiation. These reactions have been extensively studied, and it is now clear that much of the complexity is due to partitioning of the first formed intermediate into different pathways leading to a variety of products by ground state reactions. The intermediate may be stable at low temperature and may then be detected and characterized.

A few examples will illustrate the interesting transformations that can occur.

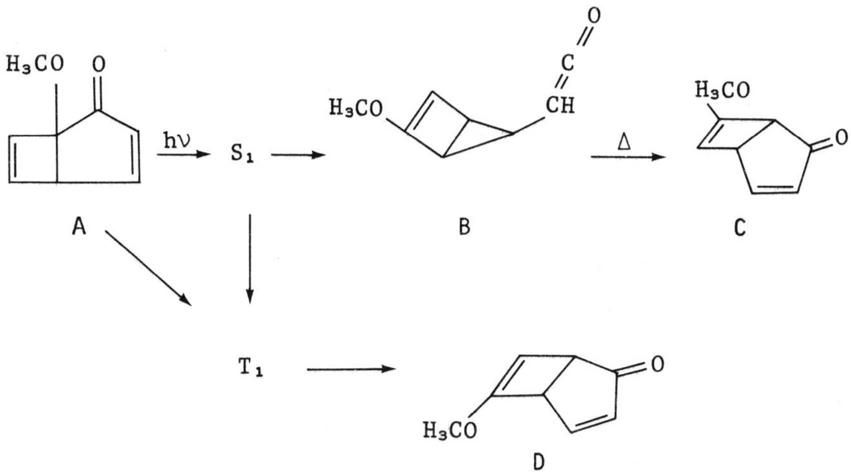

H₃CO structures... (chemical scheme)

A   →   B   →   C

T₁   →   D

Direct irradiation of the bicyclo[3.2.0]hepta-3,6-
dienone(A) gives the ketene(B) as the first detectable intermediate. The ketene rearranges thermally to the bicyclo
[3.2.0]hepta-3,6-diene(C) isomeric with the starting material. The triplet excited state (of (A)) obtained by
triplet energy transfer rearranges, presumably by a different mechanism, to the other isomer (D).

The rearrangement of 4,4-dimethyl-2-cyclohexenone involves the triplet state.

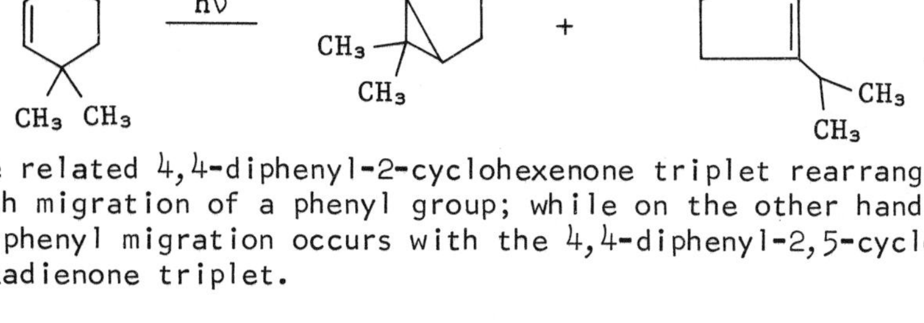

(chemical scheme)   +   (chemical scheme)

The related 4,4-diphenyl-2-cyclohexenone triplet rearranges
with migration of a phenyl group; while on the other hand
no phenyl migration occurs with the 4,4-diphenyl-2,5-cyclo-
hexadienone triplet.

179

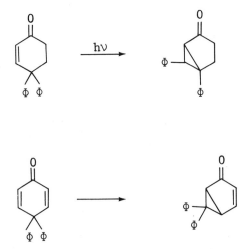

The bicyclohexenones undergo a variety of rearrangements which involve partitioning in both the excited state
and the subsequent thermal reactions.

180

The list of classes of substances undergoing rearrangement is very long and includes the following: cross-conjugated and extended dienones, polyenes, epoxyketones, enol esters, oximes, aromatic esters and ethers, N-oxides, and a wide variety of other substances.

There is no doubt that this list will be extended and that such transformations will soon be as predictable, in a qualitative sense, as are ground state reactions. Their utility in synthesis will then be fully realizable.

GENERAL REFERENCES

A. Schonberg, "Preparative Organic Photochemistry", Springer-Verlag, New York, 1968.

J.G. Calvert, J.N. Pitts, "Photochemistry", J. Wiley and Sons, Inc., New York, 1966.

Ed. R. Srinivasan and T.S. Roberts, "Organic Photochemical Synthesis, Vol. 1", Wiley-Interscience, New York, 1971.

R.O. Kan, "Organic Photochemistry", McGraw-Hill, New York, 1966.

D.C. Neckers, "Mechanistic Organic Photochemistry", Renhold Publishing Corporation, New York, 1967.

N.J. Turro, "Molecular Photochemistry", W.A. Benjamin, Inc., New York, 1965.

CHAPTER 8
SYNTHETIC APPLICATIONS
OF ORGANIC PHOTOCHEMISTRY

In the preceding Chapter several photochemical reactions were discussed, from a mechanistic point of view, which had a potential use in organic synthesis. This Chapter will present some examples of the utility of such, and other, reactions in a synthetic sequence, wherein they frequently form the key steps. Examples are selected from the fields of natural products, small ring compounds, and unusual fused ring systems, but the selection has been, perforce, arbitrary.

The use of photochemical reactions for natural product synthesis has recently been reviewed[1]. Several examples of cycloaddition steps being utilized are listed. In Corey's[2] synthesis of caryophyllene and isocaryophyllene the cyclobutane ring photochemically produced remains intact. The fact that the _trans_-ring juncture is formed preferentially in the photochemical step is particularly interesting.

------

1.   P.G. Sammes, Quart. Rev., 24, 37 (1970).

2.   E.J. Corey, R.B. Mitra and H. Uda, J. Amer. Chem. Soc., 86, 485 (1964).

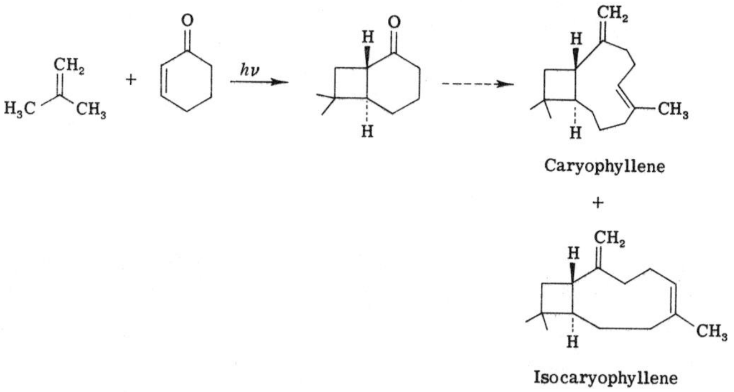

Caryophyllene

+

Isocaryophyllene

Amongst many applications one may mention the synthesis of grandisol, one of the four pheromones acting as sex-attractants for the 'boll weevil'. This is noteworthy since two photochemical syntheses have been evolved for the same product:

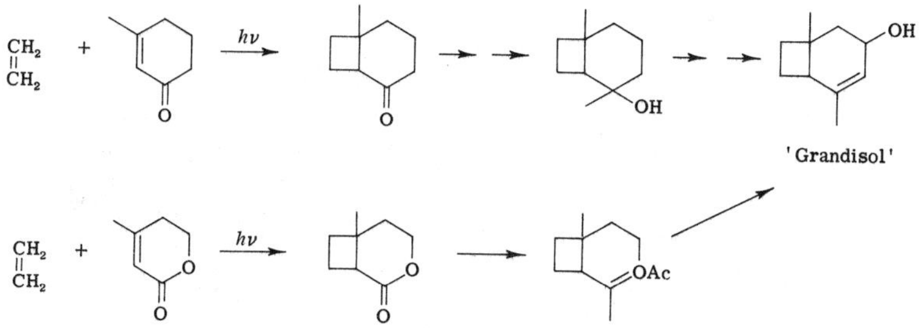

'Grandisol'

In other cases of photoannelation, the cyclobutane is ultimately cleaved. For example, the photocycloaddition of 4,4-dimethylcyclopentene to 3-methylcyclohexenone-2 is the key step in the synthesis of α-caryophyllene alcohol[3].

---

3. E.J. Corey and S. Nozol, J. Amer. Chem. Soc., 86, 1652 (1964).

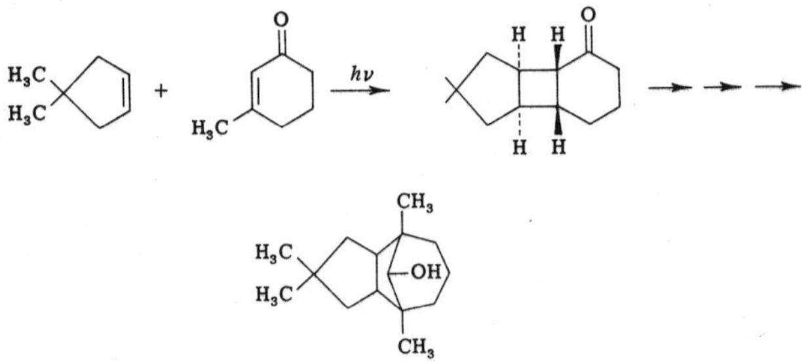

α-Caryophyllene alcohol

Natural products with the γ-tropolone moiety are available by photocycloaddition of chloroolefins to the enol acetate of 1,3-cyclopentadione. This reaction may be illustrated with the synthesis of stipitatonic acid and of γ-tropolone[4].

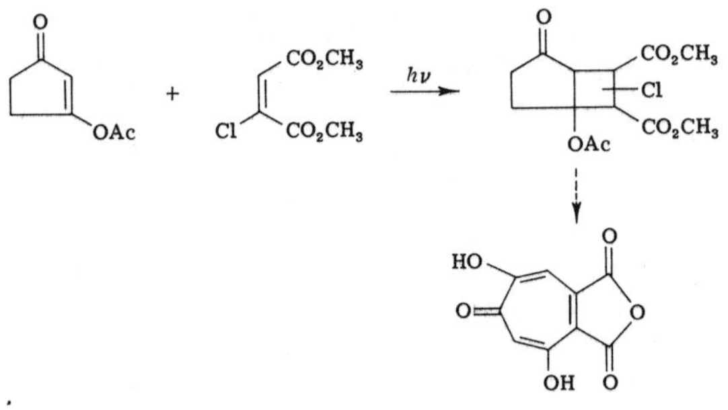

Stipitatonic acid

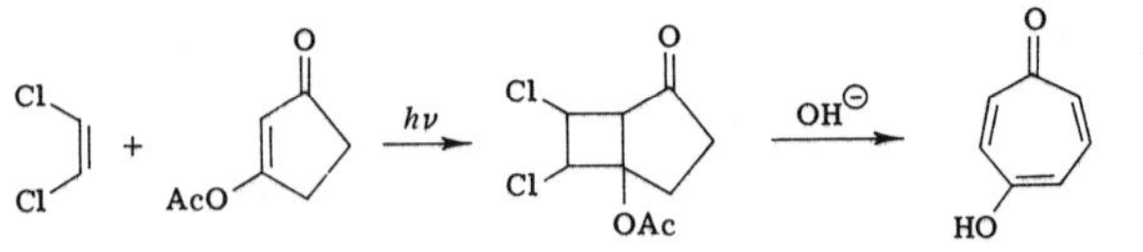

γ-Tropolone

4. B.D. Challand, H. Hikino, G. Kornis, G. Lange and P. de Mayo, _J. Org. Chem._, 34, 794 (1969).

A recent synthesis of an isomer of the naturally-occurring marasmic acid includes two such photoannelation steps (and two other photochemical reactions).

Photocyclization

There have been comparatively few examples of photo-cyclization except for what is essentially arylation; and this has been extensively employed. An early example is the use of this reaction to prepare aristolochic acid methyl ester[5].

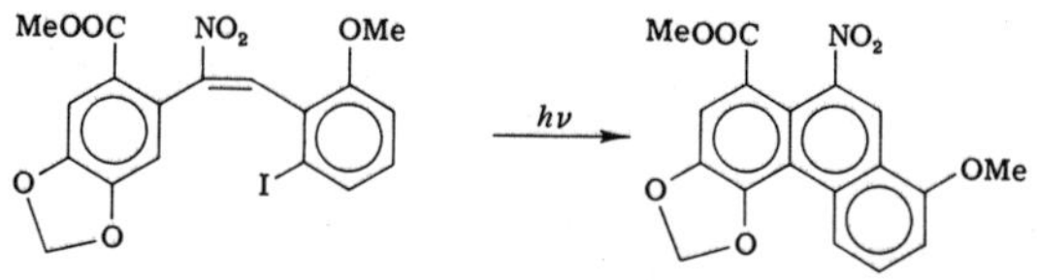

There have also been applications to aporphine synthesis:

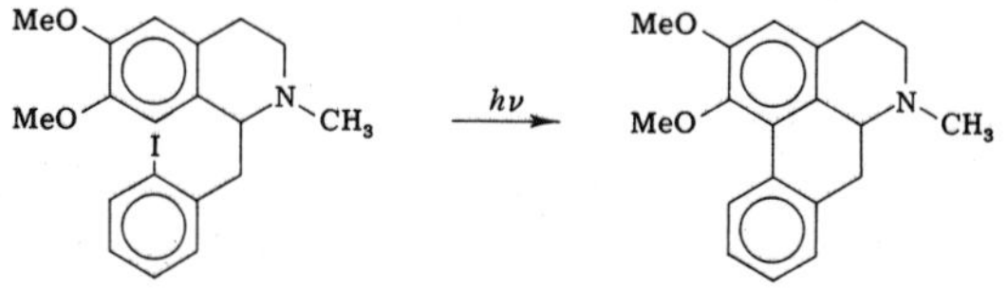

Nuciferine

---

5. S.M. Kupchan and H.C. Wormser, J. Org. Chem., 30, 3792 (1965); S.M. Kupchan, J.L. Moniot, R.M. Kanojia and J.B. O'Brien, ibid, 36, 2413 (1971).

Kametani has extensively exploited the reaction, and an example is in the following step which constitutes part of a
formal total synthesis of the crinine-type alkaloid, man-
tidine[6]:

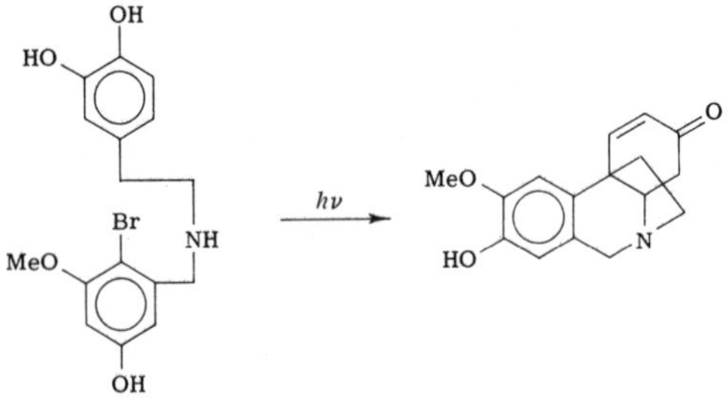

It may be noted that in certain cases photochemically-
induced electrocyclic reactions may lead to similar aromatic products. Thus the alkaloid nuciferine may also be
obtained by a synthesis in which the essential cyclization
step is as follows[7]:

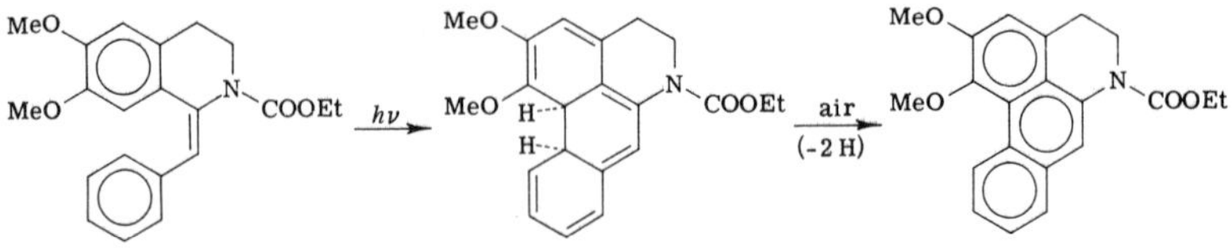

6.   T. Kametani, T. Kohno, S. Shibuya and K. Fukumoto,
      Chem. Comm., 774 (1971).

7.   M.P. Cava, S.C. Harlicek, A. Lindert and
      R.J. Spangler, Tetrahedron Letters, 2937 (1966).

Strained molecule and valence isomerism

The syntheses of a large number of strained ring molecules have been accomplished utilizing a photochemical step. An important advantage of this procedure is that low temperatures can be employed thus discouraging thermal decomposition of the product. This is well illustrated by a number of photochemically-induced valence isomerizations.

Several three-membered heterocyclic systems have been generated by irradiation of the thermally stable 1,3-dipole. The most studied of these is the nitrone-oxazirane rearrangement:

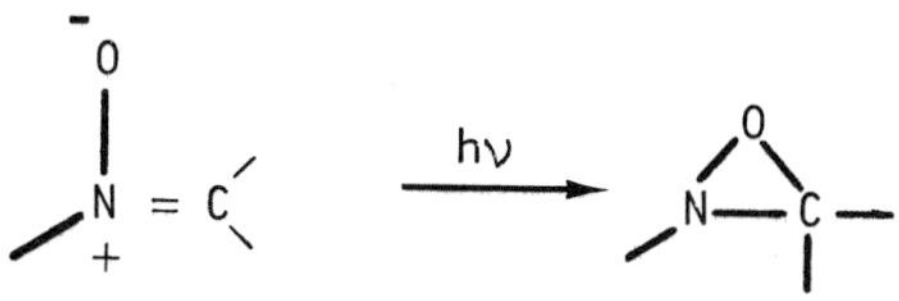

but other methods are available for the synthesis of oxaziranes. The following reaction represents the only presently known entry into the oxadiaziridine system[8]:

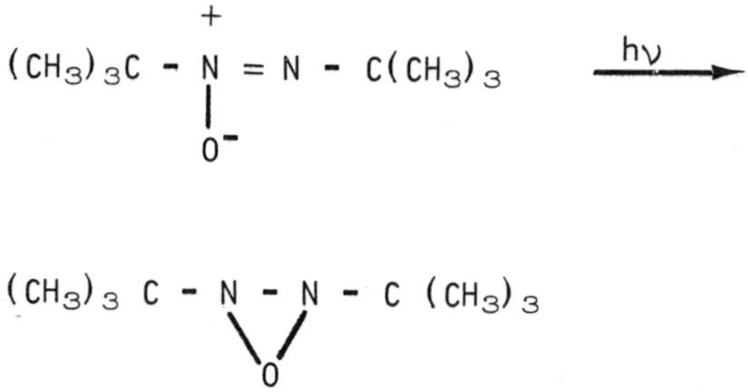

---

8.   F.D. Greene and S.S. Hecht, J. Org. Chem., 35, 2482 (1970).

   The photovalence isomerization of dienes can yield cyclobutenes or bicyclobutanes, depending upon several factors, including the configuration of the starting diene. For example, 1,3-butadiene can be converted, in a preparatively useful reaction, into cyclobutene[9], while 2,3-dimethyl-1,3-butadiene yields 1,3-dimethylbicyclobutane[10].

Also available by this reaction are unusual bi- and polycyclic systems such as bicyclo[2.1.0] pentene[11] and benzvalene[12].

---

9.   F.I. Sonntag and R. Srinivasan in "Organic Photochemical Synthesis", Ed. R. Srinivasan, Vol. 1, 1971, p. 39.

10.   J. Saltiel, R.M. Coates and W.G. Dauben, J. Amer. Chem. Soc., 88, 2745 (1966).

11.   J.I. Brauman, L.E. Ellis and E.E. van Tamelen, J. Amer. Chem. Soc., 88, 846 (1966).

12.   K.E. Wilzbach and L. Kaplan, J. Amer. Chem. Soc., 87, 4004 (1965).

Several heterocyclic systems behave similarly; thus, irradiation of 3,4-dimethyl-3-penten-2-one yields 2,3,4,4-tetramethyl-oxetene[13], and several α,β-unsaturated acids

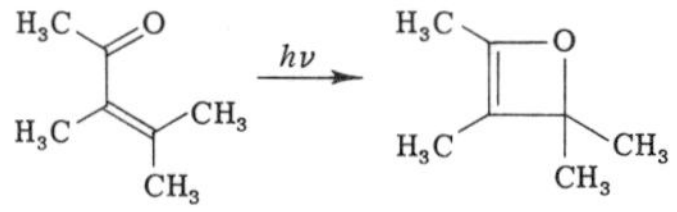

give β-lactones[14].

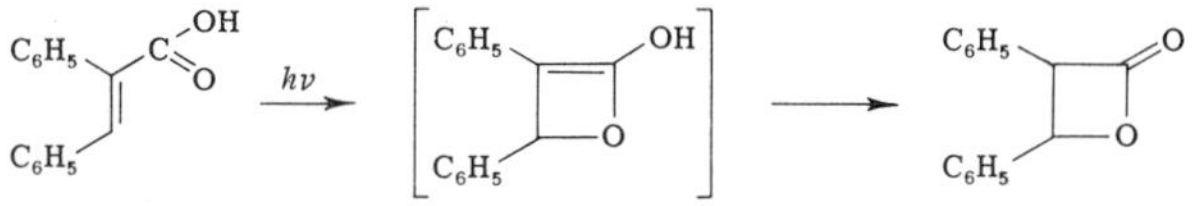

Irradiation of the diazapinone shown below leads to formation of the 1,2-diazabicyclo[3.2.0]hept-6-en-4-one[15].

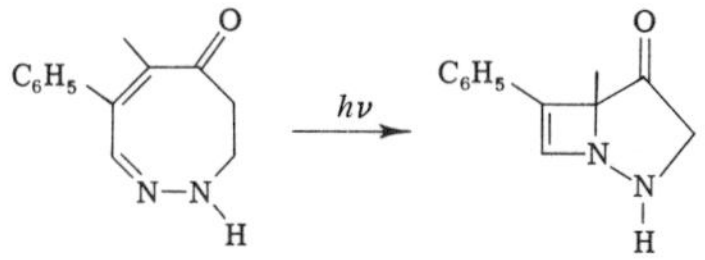

---

13.   L.E. Friedrich and G.B. Schuster, J. Amer. Chem. Soc., 91, 7204 (1969).

14.   O.L. Chapman and W.R. Adams, J. Amer. Chem. Soc., 89, 4243 (1967).

15.   W.J. Thener and J.A. Moore, Chem. Comm., 468 (1965).

A number of interesting molecules have been prepared by a sequence of which the following is typical[16]:

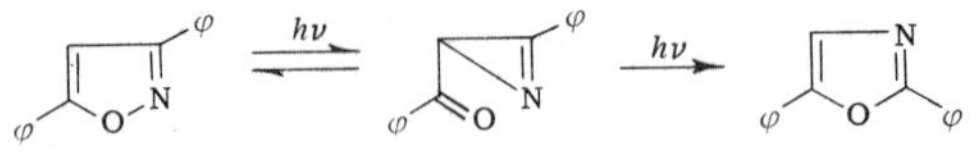

The ketocyclopropene can be isolated upon irradiation of 2,5-di-t-butylfuran[17]:

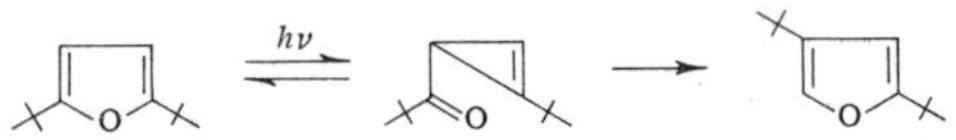

It may be mentioned here that the reverse reaction - that is, the opening of a ring system by a photochemically-induced electrocyclic reaction to give a larger ring - has also been employed synthetically[18].

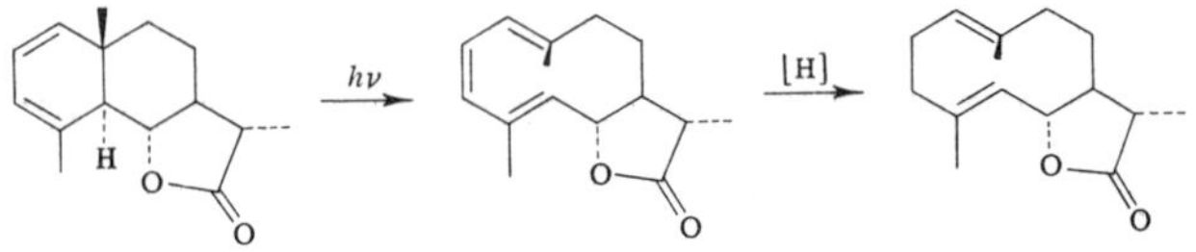

Dihydrocostunolido

16. E.F. Ullman and B. Singh, J. Amer. Chem. Soc., 88, 1844 (1966).

17. E.E. van Tamelen and T.H. Whiteside, J. Amer. Chem. Soc., 90, 3894 (1968).

18. E.J. Corey and A.G. Hortmann, J. Amer. Chem. Soc., 85, 4033 (1963).

In this connection the irradiation of 15,16-dimethyl-
pyrene is interesting. The latter is a 14π-electron system
and in agreement with Hückel's prediction, shows aromatic
properties. Irradiation, in an electrocyclic process,
cleaves the central 15,16-bond to produce the bridged stil-
bene derivative. The reaction is thermally reversible[19]:

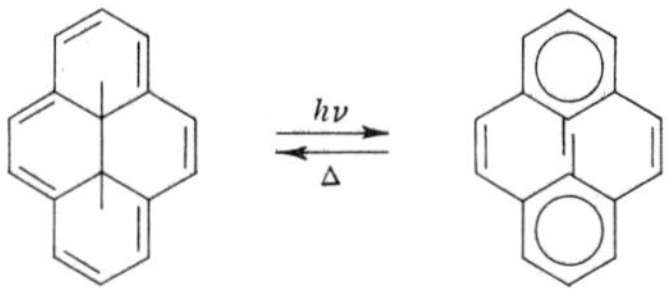

Bullvalene, the $C_{10}H_{10}$ hydrocarbon with three-fold
symmetry that has 1,209,600 equivalent structures all in-
terconverting, was first made by a photochemical reaction.
Irradiation of one of the thermal dimerization products of
cyclooctatetrene gives bullvalene in one step[20]. This is
still the best way to make this unusual hydrocarbon, although a multistep non-photochemical synthesis has since
been reported.

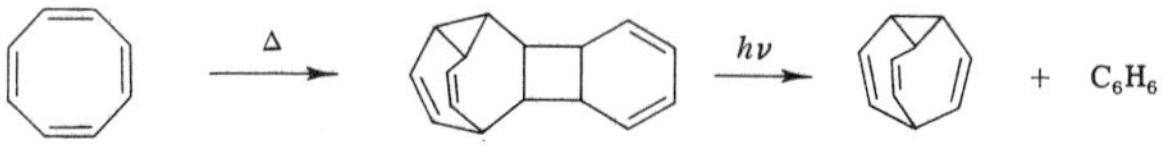

(one of the
isomeric dimers)   Bullvalene

19.   V. Boekelheide and J.B. Phillips, J. Amer. Chem.
        Soc., 85, 4033 (1963).

20.   G. Schröder, Angew. Chem. Internat. Edit., 2,
        481 (1963).

191

Cyclobutanol formation

The cyclobutanol formation which accompanies Norrish Type II cleavage of ketones possessing a δ hydrogen is itself a useful synthetic reaction and results in the generation of a strained molecule. The reaction is capable of extension to give complex fused systems, two examples of which are cited below[21].

$$C_6H_5-CO-\!\!\triangleleft\!\!\Box \xrightarrow[\text{abstraction}]{\gamma\text{-H}} C_6H_5-\underset{\cdot}{\overset{OH}{C}}-\!\!\triangleleft\!\!\overset{\cdot}{\underset{H}{\Box}} \longrightarrow \overset{C_6H_5\ \ OH}{\diamond}$$

The reaction is also able to lead to oxetanes with an exocyclic double bond[22]:

The reaction is also able to lead to oxetanes with an exocyclic double bond[22]:

---

21.   A. Padwa and E. Alexander, J. Amer. Chem. Soc., 89, 6376 (1967);

        A. Padwa and W. Eisenberg, J. Amer. Chem. Soc., 92, 2590 (1970).

22.   A. Feigenbaum and J.P. Pete, Tetrahedron Letters, 2767 (1972).

Strained fused systems may be prepared by intramolecular addition reactions. The first example was the preparation of carvone camphor by the irradiation of carvone:

This was followed some forty years later by the preparation
of the quadricyclic dicarboxylic acid below[23]:

The first synthesis of a cubane derivative also involved
such an intramolecular addition[24]:

---

23.   S.J. Cristol and R.L. Snell, _J. Amer. Chem. Soc._,
      80, 1950 (1958).

24.   P.E. Eaton and T.W. Cole, _J. Amer. Chem. Soc._,
      86, 3157 (1964).

The formation of the prismane skeleton:

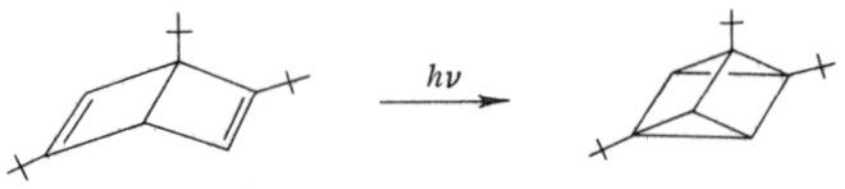

is a reaction of this type and this carbon skeleton has not been prepared by other than a photochemical route. In this case the starting diene is itself obtained by the irradiation of 1,2,4-tri-tertbutylbenzene. Such complex molecules may be formed by double addition in which the first step is intermolecular. As an example the addition of tolane to dimethylnaphthalene is cited[25]:

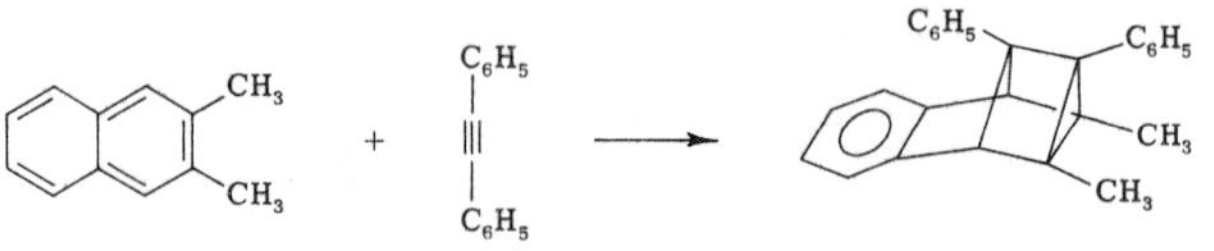

A number of other reactions, not normally considered synthetic steps, have been utilized in specific syntheses. For example the Ciamician cleavage of cyclohexanones has been used to prepare dihydronyctanthic acid[26]:

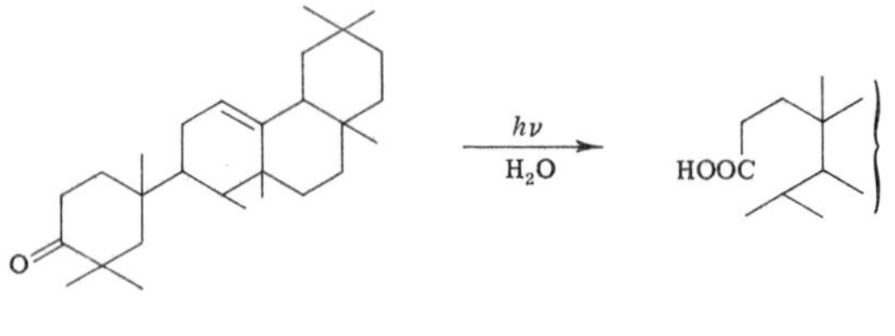

25. W.H.F. Sarse, P.J. Collin and G. Sugowdz, Tetrahedron Letters, 3373 (1965).

26. D. Arigoni, D.H.R. Barton, R. Bernasconi, C. Djerassi, J.S. Mills and R.E. Wolff, J. Chem. Soc., 1900 (1960).

The rearrangement of cross-conjugated dienones in the decalin system to substances of the perhydroazulene series is a general reaction[27]:

and it has been applied to syntheses of α-bulresene, and geigerin.

α-Bulresene   Geigerin

Amongst other reactions of synthetic utility may be mentioned the Barton nitrite photolysis and oxidation by light generated singlet oxygen. The former was introduced into the chemical scene in a spectacular way, with a very short partial synthesis of aldosterone-21-acetate. The 11-nitrite of corticosterone was photolyzed in toluene with the formation of the oxime. Hydrolysis of the latter gave the desired compound[28].

27.   D.H.R. Barton, P. de Mayo and M. Shafiq, J. Chem. Soc., 140 (1958); D.H.R. Barton, J.E.D. Levisalles and J.T. Pinney, ibid, 3472 (1962).

28.   D.H.R. Barton and J.M. Beaton, J. Amer. Chem. Soc., 83, 4083 (1961); A.L. Nussbaum and C.H. Robinson, Tetrahedron, 17, 35 (1961).

Since then the reaction has been employed many times, and is usually successful provided the stereochemical require-ment - proximity of the abstracted hydrogen - is met. A competing reaction is fragmentation.

The general principle illustrated here - that chemical activation at a distant site may follow photolytic genera-tion of a suitably positioned radical - encompasses group-ings other than nitrites. Hypochlorites and hypoiodites are other derivatives of alcohols which have been used, but derivatives of amines, amides, carboxylic acids and other functional groups have been employed. Possibly the best known is the Hofman-Loeffler-Freytag reaction which in-volves the photolysis of a protonated N-chloroamine. The formation of dihydroconnessine is illustrated[29]:

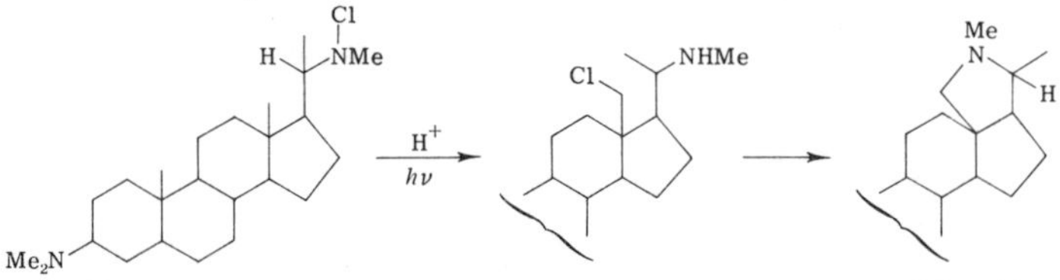

The acyl azides have been used in the diterpenoid alkaloid field to generate the required nitrogen bridge across ring A.

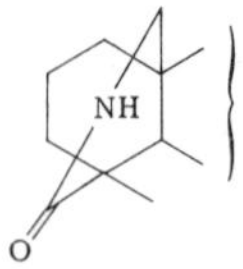

Singlet oxygen

The generation of singlet oxygen photochemically and the further reactions of this substance are not strictly within the limits of this Chapter since singlet oxygen may be generated by purely chemical means. Since, however, a number of synthetic applications are known in which the species is photochemically generated (as is virtually al-ways the case) a few examples will be cited.

---

29.   E.J. Corey and W.R. Hertler, J. Amer. Chem. Soc., 82, 1657 (1960).

There are two general types of reaction; the allylic substitution:

$$\overset{\diagdown}{C}=\overset{\underset{\displaystyle |}{H}}{\underset{\underset{\displaystyle |}{}}{C}}-\overset{\underset{\displaystyle |}{}}{C}- \quad\xrightarrow[\text{(sens)}]{O_2,\ h\nu}\quad -\overset{\overset{\displaystyle H_2O}{\underset{\displaystyle |}{}}}{C}-\overset{\underset{\displaystyle |}{}}{C}=C\overset{\diagup}{\diagdown}$$

and the addition:

where the diene usually forms part of a system in which it is fixed in the cisoid form.

Examples of the first type are the sequence:

in the synthesis of "rose oxide", the active constituent of the scent of roses, and

in which the allylic transposition produces an enol which attacks the hydroperoxide to give an epoxide.

197

Examples of the second type are:

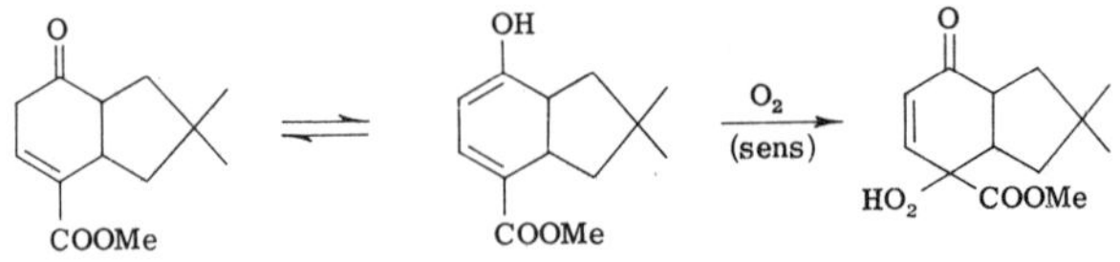

(in the synthesis of methyl isomarasmate already mentioned)
which has the appearance of an allylic substitution, the
related conversion of 7-chloroanhydrotetracycline to 7-
chloro-6-deoxy-6-hydroperoxydehydrotetracycline[30], and steps

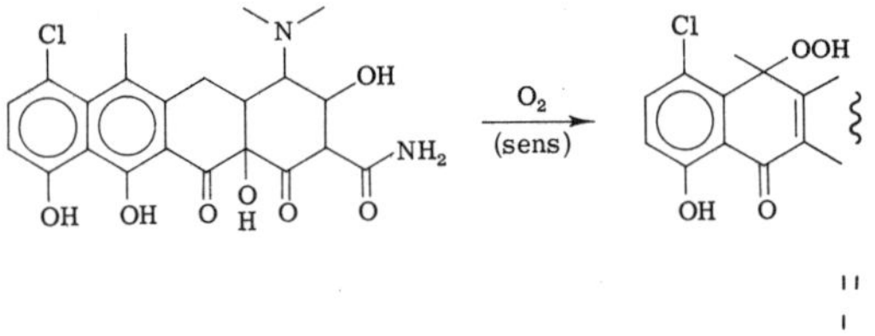

in the syntheses of cantharidin:

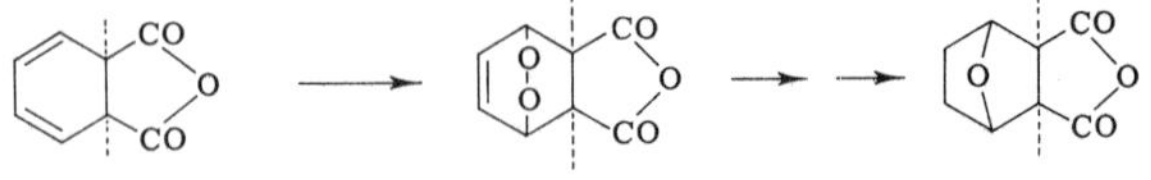

Cantharidin

30.   A.I. Scott and C.T. Bedford, *J. Amer. Chem. Soc.*,
      84, 2271 (1962).

maleimide:

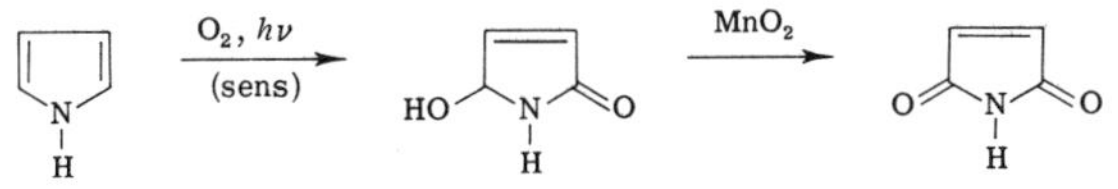

and abscisin:

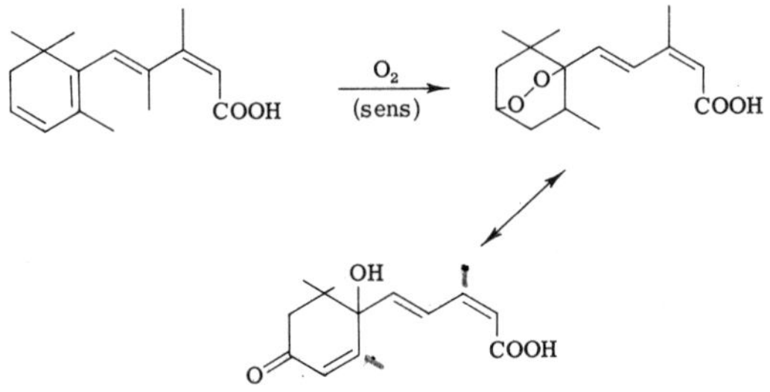

   The reaction has been extensively applied to aromatic
compounds largely in the studies of Dufraisse, Regaudy and
their school. A wide variety of aromatic substrates has
been used, including heteroaromatic compounds, and it is
not possible to summarize their results here. The reader
is referred to the succinct review in Schonberg's book[31].

Summary

   Virtually any type of chemical reaction may appear at
some time in a chemical synthesis, and photochemical reactions are no different. It is hoped that the foregoing
selection will give some idea of the variety and scope
available to chemists prepared to use the technique of
photochemistry: a range which is expanding at the present
time perhaps even faster than the already rapidly growing
chemical field.

---

31.   A. Schonberg, G.O. Schenk and O.A. Neumüller,
       "Preparative Organic Photochemistry", Springer-
       Verlag, New York, 1968, p. 389 et seq.

CHAPTER 9

PHOTOCHEMISTRY
OF THE SOLID STATE

Introduction

    The study of the photochemistry of solids is still mainly at the phenomenological stage and interpretations of experimental results are largely speculative. In contrast, the theoretical study of the solid state is well advanced and is beginning to provide quantitative explanations for many of the properties of solids - energy states, absorption spectra, luminescence, etc. - which, even if not photochemical in the strict sense of the word, are at least plainly related to photochemical events. It therefore seems appropriate in this chapter to give an introductory account of those parts of solid state theory which seem to be most essential for future quantitative understanding of photochemical processes in solids. In the last section, three examples of solid state photochemical reactions will be discussed in terms of the concepts developed in sections (1) through (3).

    It will be seen that some simple photochemical processes such as dimerizations are governed by the arrangement of the molecules in the crystal, but when the geometrically favored product cannot form for steric reasons then reaction occurs in defective regions of the crystal where the reverse of the normal arrangement of the molecules obtains. This is but one example - and there are many - of the importance of crystal imperfections in solid state reactions. Again, to elucidate the mechanism of an electron transfer reaction we should understand something about electron energy levels in crystals and know what species are formed as a result of electron excitations. Such understanding is dependent upon realizing the all-pervading influence of translational symmetry in crystals. A large part of the following chapter on photochromism will also depend on many of the concepts developed in this one, particularly energy bands, impurity levels and color centers.

1.   ENERGY STATES IN CRYSTALS

Translational symmetry

The fundamental property which distinguishes a crystalline solid is its translational symmetry. A perfect solid is infinite in extent and consists of identical clusters of atoms (or molecules or ions) grouped around a set of fixed points in space called lattice points, such that the environment of every lattice point looks the same. The lattice points are defined by vectors of the form

$$\underset{\sim}{\ell} = \ell_1 \underset{\sim}{a_1} + \ell_2 \underset{\sim}{a_2} + \ell_3 \underset{\sim}{a_3} \qquad (1)$$

where $\underset{\sim}{a_1}$, $\underset{\sim}{a_2}$, $\underset{\sim}{a_3}$ are three non-collinear primitive translation vectors and ℓ_1, ℓ_2, ℓ_3 are integers. The identical set of atoms grouped around each lattice point is called the basis and the combination of lattice + basis forms a crystal structure. The basis is specified by the types of atoms and their distances from the lattice point with which they are associated. A simple two-dimensional (2-D) example is given in Figure 1. The lattice points are at the positions $\ell_1 \underset{\sim}{a_1} + \ell_2 \underset{\sim}{a_2}$ and the atom positions are specified by the vectors

$$\underset{\sim}{x}_A = 0$$

$$\underset{\sim}{x}_B = \tfrac{1}{2} \underset{\sim}{a_1} + \tfrac{1}{2} \underset{\sim}{a_2} \qquad (2)$$

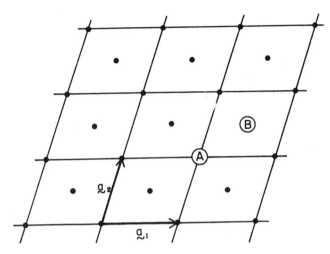

Figure 1.   A simple example of a two-dimensional lattice
             showing the primitive translational vectors
             a_1 and a_2 that generate the lattice.   The
             basis consists of atoms of type A situated
             at the lattice points and atoms of type B
             located at the centers of the parallelograms
             formed by joining the lattice points.

     In Figure 1 the angle θ between a_1 and a_2 is neither
$60°$ nor $90°$.   If it had either of these values the lattice
would clearly have more symmetry than the one shown; the
overall symmetry depends on the relative length of a_1 and
a_2 and on θ.   There are five distinguishable lattices in
2-D and fourteen in 3-D.   Such lattices are called the
Bravais lattices and they are enumerated in almost every
standard text[1].

     A little thought shows that a certain arbitrariness
exists in the choice of a_1, a_2 and a_3.   The parallelopiped
bounded by a_1, a_2, a_3 is called the <u>unit cell</u> of the
lattice.   If it contains a single lattice point it is called

----

1.   See, for example, F.C. Phillips, "An Introduction to
      Crystallography", 3rd ed., p. 226, Longmans,
      London, 1962.

the primitive cell of the lattice.   It is often convenient
to regard some of the Bravais lattices as a combination of
a simpler Bravais lattice and a basis; the usual reason is
to achieve greater symmetry.   The price one has to pay for
this gain in simplicity is that the unit cell is no longer
primitive but contains more than one lattice point.   The
primitive cell does not always display the representative
symmetry of the lattice.   One type of unit cell of great
importance which is primitive is the Wigner-Seitz cell;
this is defined as the polydedron resulting from the intersection of a set of planes that bisect perpendicularly the
lines joining a lattice point to all its near neighbors.

The reciprocal lattice

This is a concept of central importance in describing
both lattice vibrations and electronic states in crystals.
The reciprocal lattice is the lattice based on the vectors
b_1, b_2, b_3 which satisfy the relations[2]

$$\underset{\sim}{b}_i \cdot \underset{\sim}{a}_j \;=\; 2\pi\delta_{ij} \qquad i,j \;=\; 1,2,3 \qquad (3)$$

For example, the reciprocal lattice of a fcc lattice is
bcc, and vice versa, the cube edge in the reciprocal lattice being $4\pi/a$ where a is the cube edge in the direct lattice.   The importance of the reciprocal lattice concept lies
in the fact that both electronic state functions and lattice
vibrations are described in terms of wave vectors which are
vectors in the space of the reciprocal lattice.

---

2. δ_{ij} is the Kronecker delta symbol; it is equal to
zero when $i \neq j$ and to unity when $i = j$.   Note that
in crystallography the reciprocal lattice is often
defined without the factor 2π.   Also, in crystallography the primitive lattice translational vectors
are usually represented by $\underset{\sim}{a}$, $\underset{\sim}{b}$, $\underset{\sim}{c}$, instead of $\underset{\sim}{a}_1$,
$\underset{\sim}{a}_2$, $\underset{\sim}{a}_3$.

Space groups

The operation t of displacing a lattice point by the vector ℓ [**Eq.**. (1)] is clearly a symmetry operation, since the environment of any lattice point looks the same after t as it did before the operation was carried out. The totality of all such operations t called the primitive translations (i.e. for all possible values of the integers ℓ_1, ℓ_2, ℓ_3) is the translation group.

A space group is a collection of symmetry operations that leave a lattice invariant. Such space symmetry operations consist of a combination of a point symmetry operation (rotation) with a translation. If the rotation operations form a point group then the space symmetry operations form a group (in a mathematical sense) called a space group. The general space group operation need not involve a primitive translation. But if it is a primitive translation then the rotational symmetry operation must leave the lattice invariant: this fact is of great importance for it is this property of a lattice which limits the number of space lattices to the 14 Bravais lattices. The general space group operation may be regarded as the successive application of a primitive translation and the combination of a point symmetry operation and non-primitive translation. The symmetry elements corresponding to such space operations include rotation axes, rotoinversion axes, screw axes and glide planes. There are but 32 point groups, called the crystallographic point groups, that are compatible with 3-D translational symmetry. The combination of the crystallographic point groups with the primitive and nonprimitive translations results in the 230 space groups. Space groups are identified by symmetry symbols. For example, the symbol $P2_12_12_1$ indicates the combination of three mutually perpendicular two-fold screw axes with the translations of a primitive orthorhombic lattice.

Types of bonding

The properties of crystalline solids depend to considerable extent upon the nature of the chemical binding forces between the structural units (atoms, molecules or ions) of which the crystal is composed. In order to understand the nature of these binding forces the electronic structure of the atoms making up the crystal must be considered.

Neon, with atomic number $Z = 10$ is monatomic with completely filled sub-shells. The intermolecular forces between neon atoms are relatively weak and the cohesive energy of solid neon correspondingly low (0.45 kcal/mole). Sodium ($Z = 11$) has a single 3s electron and solid sodium may be regarded as a collection of positive ions, associated with a lattice, through which circulate a number of 'free' electrons, one for each sodium atom. The primary contribution to the binding energy arises from the attraction between the positive ions and the electrons but this is offset to a significant extent by the contribution from the average kinetic energy of the electrons. This is a consequence of the Pauli Exclusion Principle which allows only two electrons into the lowest energy level and thus forces all the remainder to go into states which are energetically less favorable. The resulting binding energy is about 26 kcal/mole.

Chlorine ($Z = 17$) has one valence electron missing from the 3p shell. A chlorine atom thus forms a homopolar bond with a second atom to form an electronically saturated molecule. The forces between molecules with completely filled sub-shells are weak. When sodium and chlorine are brought together transfer of the 3s electron from Na to Cl results in Na^+ and Cl^- ions. Despite the fact that each now has complete valence shells, there are strong coulombic forces between the oppositely charged ions and the cohesive energy of NaCl is high, 183 kcal/mole. The coulombic contribution to the cohesive energy is offset to an extent of around 10% by a repulsive contribution arising from the overlap of the ions with closed shell configurations.

The fourth type of bonding that occurs in crystalline solids is exemplified by diamond ($Z = 6$) in which four directed homopolar bonds are formed with neighboring atoms giving a very strong giant molecule. These four principal types of crystalline solids are called molecular crystals, metals, ionic crystals and valence crystals.

Free electron theory of metals

In 1900 Drude suggested that the valence electrons in a metal are able to move freely within the boundaries of the metal, constituting what is often called the 'electron gas', while the rest of the atoms - the positively charged ion cores - occupy the crystal lattice sites. This model obviously accounts for the good electronic conductivity of metals. To account for the finite resistivity of metals it was assumed that the electrons lose kinetic energy by colliding with the ion cores. But such a model would predict an electronic contribution to the molar heat capacity of $\frac{3}{2}R$ for each free electron whereas the total heat capacity at around room temperature is usually quite close to the value 3R expected from the vibrations of the atoms. The predicted value for the paramagnetic susceptibility of the conduction electrons is also much too high. Both these difficulties are removed by the application of quantum mechanics to this problem.

Each of the N electrons is assumed to move in an averaged field due to all the positively charged ion cores and the N-1 other electrons. The potential energy V resulting from the interaction with this field is constant within the metal and rises discontinuously to a high value at the boundary of the metal. The solutions to the Schrödinger equation in this simple situation [setting the (arbitrary) constant value of V equal to zero] are

$$\Psi(\underset{\sim}{r}, t) = V^{-\frac{1}{2}} \exp\{i(\underset{\sim}{k} \cdot \underset{\sim}{r} - \omega t)\} \qquad (4)$$

$$E = \frac{\hbar^2 k^2}{2m} \qquad (5)$$

Equation (4) describes a travelling wave with $\omega = 2\pi\nu = E/\hbar$; E is the total (= kinetic) energy of the electron of mass m; $\underset{\sim}{r}$ describes the position of the electron and V is the volume of the crystal. $\underset{\sim}{k}$ is

206

the wave vector which describes the direction of
propagation of the wave and also its wavelength
since $|\underset{\sim}{k}| = 2\pi/\lambda$. Note that the energy is pro-
portional to k^2. The periodicity of the crystal
does not enter into the calculation since $V = 0$.
Quantization is effected as usual by the boundary
conditions which are of the form

$$\Psi(x + L_1) \quad = \quad \Psi(x) \tag{6}$$

where L_1 is the linear dimension of the crystal
in the x direction. This limits the values of
the components of $\underset{\sim}{k}$ to

$$k_1 = 2\pi n_1/L_1, \quad k_2 = 2\pi n_2/L_2, \quad k_3 = 2\pi n_3/L_3 \tag{7}$$

so that, for a cube $L_1 = L_2 = L_3 = L$,

$$\underset{\sim}{k} \quad = \quad 2\pi \underset{\sim}{n}/L \tag{8}$$

where $\underset{\sim}{n}$ denotes the set of integers n_1, n_2, n_3.

The energy is thus quantized to the allowed values

$$E \quad = \quad \frac{n^2 h^2}{2mV^{\frac{2}{3}}} \tag{9}$$

In accordance with the Pauli Exclusion Principle the
electrons occupy these energy levels two at a time until
they are all accounted for. The requisite number of electrons can be accommodated within an energy span of a few
electron volts because the energy levels are very closely
spaced and because the degeneracy increases rapidly with n.

Density of states and fermi energy

The most important property of these energy states is their density i.e. the number of states between E and E + dE, written $g(E)\,dE$. It follows solely from the boundary conditions that the density of states is given by

$$g(E) \quad = \quad \frac{4\pi V}{h^3} \ (2m)^{\frac{3}{2}} \ E^{\frac{1}{2}} \tag{10}$$

This function is shown in Figure 2. At 0 K the electrons all go into the lowest possible states consistent with the Pauli principle; the highest occupied state is designated E_F^o. At a finite temperature the electron distribution is governed by Fermi-Dirac statistics so that the probability of a given state E being occupied is

$$f(E) \quad = \quad [1 + \exp\{(E - \mu)/kT\}]^{-1} \tag{11}$$

μ is the chemical potential of an electron in the metal. At all T > 0 K, $f(E) = \frac{1}{2}$ at $E = \mu$. The value of E defined by $f(E) = \frac{1}{2}$ is called the _Fermi energy_ and denoted by the symbol E_F. The density of filled states for T > 0 K is also shown in Figure 2.

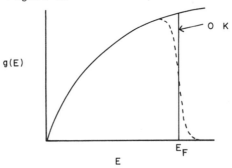

Figure 2.   Density of states $g(E)$ at 0 K. The dotted line shows the density of filled states for T > 0 K. The Fermi energy shifts very slightly with temperature.

Effect of periodic potential

The free electron theory describes the properties of
simple metals (eg. Na, Ag) quite well. It is quite unable
to account for the existence of insulators or semi-conduc-
tors. Whatever the details of the potential energy in a
real crystal (set = 0 in the FE approximation) it must
manifest the periodicity of the lattice. Superimposing a
periodic potential upon the constant (zero) potential of
the FE theory has two results. Firstly, the plane waves (4)
become modulated with the periodicity of the lattice.
Secondly, the energy is no longer a continuous function of
k but the energy states become broken up into bands (Figure
3). The bands are separated by forbidden energy gaps in
which there are no energy states in a perfect (infinite)
crystal.

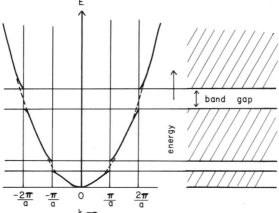

Figure 3.   Dependence of energy on k in a specific
                direction in the reciprocal lattice for a
                model periodic potential. Only certain
                energies are allowed leading to bands of
                allowed energy states, separated by gaps, as
                shown on the right hand side of the figure.
                The dotted lines show the E(k) parabola of
                the free-electron theory.

For each direction of the vector k in the reciprocal
lattice the discontinuities in E occur at certain values of
k and the loci of the points at which these discontinuities
occur generate polyhedra in k-space called Brillouin Zones.
The first Brillouin Zone is just the Wigner-Seitz cell of

the reciprocal lattice. The shapes of the Brillouin Zones are determined solely by the crystal lattice but the determination of the energy (as a function of k) at each point in the zone requires the solution of the Schrödinger equation. In presenting the results of such a calculation it is usual to display E as a function of k along some of the principal directions (of high symmetry) in the Brillouin Zone. Since in any one of these directions the discontinuities occur at regular multiples of k (see Figure 3) and since moreover E is an even function of k, it is usual to plot E as a multivalued function of k in a single half-zone. This is called the reduced zone scheme.

Effective mass; positive holes

A simple differentiation of E in equation (5) gives

$$\frac{1}{m} = \frac{1}{\hbar^2} \frac{d^2E}{dk^2} \tag{12}$$

This shows that the curvature of the parabolic function E(k) is related to the mass of the electron. Corrections for near free-electron behavior can be accounted for by using the FE density of states (10) but replacing m by an effective mass

$$m_e = \frac{\hbar^2}{d^2E/dk^2} \tag{13}$$

where the denominator is the curvature of the actual (no longer parabolic) function E(k). Clearly the effective mass may vary with E, and with the direction in k space, but over a restricted range of energies in a specific direction in the Brillouin Zone it may be treated as a constant. Numerical values for effective mass are usually quoted as the ratio m_e/m.

Inspection of Figure 3 shows that the curvature of E(k) becomes negative near the top of a band! However, an empty state in an otherwise filled band behaves as if it

210

were a single particle of positive charge e and effective mass

$$m_h = - \hbar^2 \left(\frac{d^2E}{dk^2} \right)^{-1} \tag{14}$$

Such a vacant state in an otherwise filled band is called a positive hole.

Metals, insulators and semiconductors

The classification of solids into metals, insulators and semiconductors is based on their electronic conductivity. Metals are good conductors at ordinary temperatures and their conductivity decreases with rise in temperature. Insulators do not exhibit electronic conductivity at any temperature (excepting, possibly, photoconductivity). Semiconductors are insulators at low temperatures but their conductivity increases with rising temperature.

The existence of metals, insulators and semiconductors finds a ready explanation in terms of the band model. If a substance has a band which is approximately half full (as in the alkali metals of group I) or a full band which overlaps an empty band (as in the alkaline earth metals of group II) then it behaves as a metal because there are abundant energy states available as in the free electron model. But if the material has a band which is completely filled with valence electrons (valence band) and this is separated by a large energy gap from a completely empty band, the substance in question is an insulator as far as electronic conduction is concerned. (Such substances may, as we shall see, exhibit ionic conductivity.) If the valence band is separated from the conduction band by a small gap (order of magnitude 1 eV) the electrons may be excited thermally from the valence band to the conduction band giving rise to semiconduction. Both electrons in the conduction band and the holes in the valence band contribute to the semiconductivity.

This type of semiconductivity is a property of the pure material which is therefore called an intrinsic semi-

211

conductor. Extrinsic semiconductivity is due to the presence of impurities which provide either extra electrons in energy levels below the conduction band, or vacant states above the valence band. In the first case (n type semiconductors) electrons are excited from the impurity levels into the conduction band; in the second instance p type semiconductors) the conductivity is due to holes in the valence band generated when electrons are excited into the vacant impurity levels.

Both n and p type semiconductors have energy states within the forbidden gap caused by impurities. Such energy states also arise from crystal imperfections and from the surface of the crystal.

2.   IMPERFECTIONS IN CRYSTALS

All real crystals contain imperfections of one sort or another. These imperfections may be classified as point defects, linear defects, planar defects, and complex defects which involve one or more defects in association. The phenomena of diffusion and ionic conductivity in crystals indicate the existence of point defects. As the temperature is raised certain atoms or ions may leave their normal lattice sites. If an ion goes into an interstitial position the interstitial together with the vacancy form a Frenkel defect. If a pair of ions of opposite sign take up new positions on the surface of the crystal then the resulting pair of vacancies is called a Schottky defect. Such processes require energy but there is a compensating gain in entropy and so at any particular temperature the crystal contains the appropriate number of defects which minimize its free energy. Various types of simple and complex point defects are shown in Figure 4. Point defects (even vacant sites) can be regarded as thermodynamic species possessing entropy, chemical potential, and the other thermodynamic properties. Their formation can be described in terms of quasi-chemical relations such as

$$Ag^+ \rightarrow Ag_i^+ + \boxed{+} \qquad (15)$$

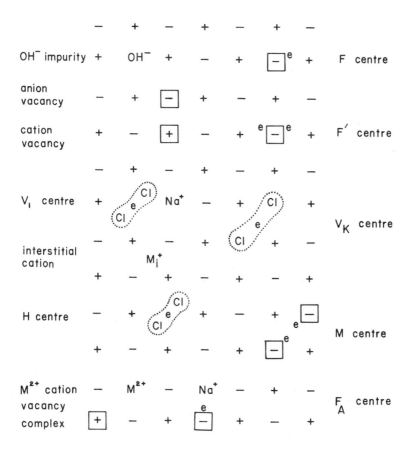

Figure 4.   Schematic representation of various simple
and complex point defects in an ionic
crystal. Interstitial cations M_i^+ occur in
AgCl and AgBr. All the other defects shown
are known to occur in KCl and many analogous
ones will occur in other crystals.

213

which describes the formation of Frenkel defects on the cation sub-lattice of silver bromide or silver chloride. The subscript i denotes that the normal lattice cation Ag^+ has been displaced into an interstitial position while the symbol $\boxed{+}$ denotes the cation vacancy left behind. The defect equilibria corresponding to these quasi-chemical processes can be written down immediately; for example the site fraction of cation vacancies c_1 and interstitials c_i are related by the equation

$$c_i c_1 \ = \ K_F \ = \ 2 \exp(-g_F/kT) \qquad (16)$$

K_F is the equilibrium constant for the thermodynamic process (15) and is connected with the Gibbs free energy of formation of the Frenkel defect pair g_F by the usual relation. The factor 2 comes from the fact that in the AgCl structure there are two interstitial positions for every cation site. A similar type of relation would hold for Schottky defects.

All real crystals contain impurities and many materials contain impurities that have been added intentionally to endow the substance with certain desirable properties. This is known as doping. A common example of an impurity point defect is an OH^- substituting for a normal lattice anion in KCl. If the substitutional impurity has a valency different from that of the normal lattice ion then charge compensation requires that a vacancy exist somewhere else on the same sub-lattice. For example, if an AgCl crystal contains $CdCl_2$ in solution, the Cd^{2+} ions are accommodated on the normal cation sites and for each Cd^{2+} there is also a vacant cation site. Now a Cd^{2+} carries one more positive charge (i.e. 2) than the normal lattice cation and the vacancy one less positive charge (i.e. 0) than the normal lattice cation, so that an electrostatic interaction exists between them resulting in a more stable configuration when the Cd^{2+} ion and the vacancy are on adjacent cation sites. This is called a complex and resembles ion pairing in solution. There is also an interaction between defects which are not on nearest neighbor sites and such interactions can be described reasonably well by the Debye Hückel theory of ion interactions in dilute electrolyte solutions.

Aliovalent ions introduced into semiconductors affect the concentrations of free electrons and holes. For example if NiO is doped with Li^+ ions then charge conservation is preserved by an equal number of Ni^{2+} ions changing their valence to the trivalent state. The Ni^{3+} ions provide impurity levels just above the valence band. Charge transport occurs by excitation of an electron into these levels leaving a positive hole in the valence band. However, holes in NiO have a very low mobility showing that they are easily trapped and require activation energy to jump from one site to the next. Thus NiO is sometimes called a 'hopping semiconductor'.

Color centers

When alkali halide crystals are subjected to ionizing radiation they adopt a characteristic color, blue for KBr, magenta for KCl, etc. This color is due to the presence of a bell-shaped absorption band peaking at around 540 nm in KCl. The centers responsible for this absorption were investigated intensively in Göttingen in the 1930's by Hilsch and Pohl, who termed them 'Farbzentren'. Whole families of color centers have now been characterized and the one originally investigated at Göttingen is called the F center. It consists of an electron trapped at an anion vacancy. A vacant anion site carries an effective positive charge because there is a missing negative charge at that point in an otherwise uniform dielectric; thus it attracts the electron rather as does the nucleus of a hydrogen atom giving rise to a series of bound states. The observed absorption band is due to the transition of this electron from 1s-like ground state to a 2p-like excited state. Irradiation in the F band at low temperatures produces F'centers which consist of two electrons trapped at a single anion vacancy. Irradiation in the F band at higher temperatures produces new absorption bands on the low energy side of the F band. These are due to M centers which consist of two F centers on nearest neighbor sites ($\equiv F_2$), R centers ($\equiv F_3$) and higher aggregates. Several of these types of centers are shown in Figure 4.

If an alkali halide crystal is heated in the vapor of the corresponding metal, say NaCl in Na vapor it becomes non-stoichiometric. The Na atoms enter the crystal as Na^+ ions and F centers are formed to provide conservation of charge and lattice sites. If the additively-colored NaCl is heated to 400°C without irradiation colloidal particles of Na form. These induce a blue color due to the scattering of light from the small particles which are 10-50 Å in diameter. By heating to higher temperatures still the colloids can be dispersed reforming F centers.

When an alkali halide crystal is exposed to ionizing radiation in addition to the prominent F band there are formed a number of overlapping absorption bands on its high energy side. These bands are associated with various hole centers. At liquid helium temperatures the most important of these is the H center, which consists of an X_2^- molecular ion occupying a single anion site. At higher temperatures (liquid nitrogen) the most important are the V_K center which is an X_2^- molecular ion (equivalent to a self-trapped hole) and the V_1 center which, in KCl, is an H center adjacent to a Na^+ cationic impurity. A summary of the principal types of color centers is given in Table 1 and many of these centers are also illustrated in Figure 4.

The mechanism of formation of these centers is of central importance in understanding radiation damage in solids. It has been a subject of much dispute over a number of years and even now is not understood in its entirety. The Pooley mechanism envisages (i) the formation of electrons and holes by the radiation; (ii) the trapping of a hole at a normal anion site to give a V_K center; (iii) the recombination of electrons with V_K centers to give a highly unstable $[Cl_2^=]^*$ which decays to normal Cl^- anions with the liberation of kinetic energy; (iv) use of this kinetic energy to form an interstitial Cl_i^- by a $\langle 110 \rangle$ re-

TABLE 1

Principal types of color centers
in alkali halide crystals

(see also Figure 4)

Center	Structure
F	electron trapped at an anion vacancy
F'	two electrons trapped at a single anion vacancy
M	two F centers on adjacent sites, aligned in $\langle 110 \rangle$ on $\{100\}$ type planes
R	three F centers forming an equilateral triangle in $\{111\}$ planes
F_A	F center adjacent to an Na^+ or Li^+ impurity cation in KX
H	X_2^- molecular ion on one anion site
V_K	X_2^- molecular ion on two (adjacent) anion sites
V_1	H center adjacent to an Na^+ cation (in KCl)

placement sequence[3], leaving an anion vacancy separated from the interstitial by several lattice sites so that re-combination is unlikely: (v) trapping of electrons at the

---

3.　Any particular direction in a crystal lattice can be described uniquely by specifying the components along the a_1, a_2, a_3 axes of a vector along this direction. The components are always given as the smallest set of integers that result after removal of fractions. They are enclosed by square brackets to indicate a specific direction and by angle brackets to indicate a set of equivalent directions.

anion vacancies to give F centers and of the holes elsewhere to give V_K centers. In favor of this mechanism is the observation of recombination luminescence at low temperatures and the fact that F center production falls off as the luminescence increases. This would seem to indicate that the necessary kinetic energy to form the interstitial comes from a radiationless transition, namely (iii) above. The chief difficulty is that the energy available appears to be lower than that calculated to initiate a <110> replacement sequence.

Dislocations

The yield stress of real crystals is several orders of magnitude less than calculated values. The reason for this is that they contain linear imperfections called dislocations. There are two basic kinds of dislocations, edge and screw dislocations (Figures 5 and 6). A screw dislocation is parallel to the direction of slip and an edge dislocation is perpendicular to the direction of slip. In general any particular dislocation may be resolved into edge and screw components. The characteristic quantity describing a dislocation is its Burgers vector b. (Figure 5 and 6) Edge dislocations can move easily in~their slip plane thus accounting for the relative ease of plastic flow. Dislocations have a high strain energy because they involve a local distortion of the lattice. They are thus not equilibrium imperfections but are formed during crystal growth. They are also generated during cold working. Dislocations may be observed directly by electron microscopy, by decoration with precipitates (eg. Ag in KCl) or by etch pits formed because the crystal will dissolve first at the points of emergence of dislocations.

Actual crystals consists of mosaic blocks, containing relatively few dislocations, separated from another by regions containing a high density of dislocations called grain boundaries. The formation of grain boundaries between misaligned blocks of good crystal is illustrated in Figures 7 and 8. Apart from their influence on the mechanical

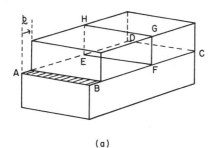

(a)

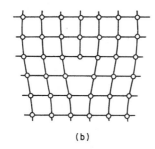

(b)

Figure 5. (a) EF is an edge dislocation lying in the slip plane ABCD perpendicular to the direction of slip. EFGH is the extra half plane of atoms above the slip plane. b is the Burgers vector.

(b) Atomic disregistry in the neighborhood of an edge dislocation is shown by a section normal to EF.

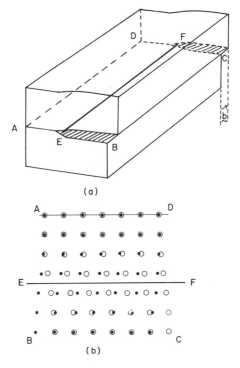

(a)

(b)

Figure 6.   (a)   In a screw dislocation the dislocation
                  line EF is parallel to the direction of
                  slip denoted by b.
                                    ~

            (b)   Atomic disregistry in the neighborhood
                  of a screw dislocation. Atoms on the
                  lower plane are represented by • and
                  those on the upper plane by ɔ . At
                  the dislocation line the slip is half
                  one lattice parameter.

properties of solids dislocations are of importance in chemistry because they can act as sites for the initiation of solid state and gas + solid reactions.

Figure 7.   Structure of a simple <u>tilt</u> <u>boundary</u>. The crystallographic planes in two adjacent blocks are tilted by an angle of a few degrees with respect to one another. The result is a series of parallel edge dislocations spaced at equal intervals along the boundary. Note that the two blocks can be brought into the same orientation by rotation about an axis that lies in the boundary; this is characteristic of a tilt boundary.

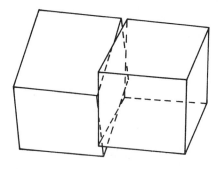

Figure 8.   If the two blocks can be brought into the
same orientation by rotation about an axis
that is normal to the boundary, it is a
simple <u>twist boundary</u>. In general, a
boundary may contain tilt and twist components. A pure twist boundary consists
of a crossed grid of screw dislocations.

Lattice vibrations

The atoms in a crystal do not occupy precisely the
positions decreed by the symmetry of the lattice but are
in a constant state of vibration about their equilibrium
positions. Their complex motion can be resolved into 3N
normal modes of vibration and the study of the frequencies
of these normal modes is an important part of the subject
of lattice dynamics. The vibrations may be classified as
longitudinal vibrations, in which the atoms move closer together or further apart along a line, or transverse vibrations, in which the atoms move perpendicularly to a line of
atoms. Adjacent atoms may be moving together essentially
in phase (the acoustic modes) or out of phase (optical
modes). Information about the lattice vibration frequencies is summarized in two kinds of figures: dispersion
curves, which are plots of frequency ω versus wave vector $\underset{\sim}{k}$,
and frequency distributions which give the fractional number
of normal modes with frequencies between ω and $\omega + d\omega$.

The vibrational motion of atoms is quantized and a quantized unit of vibrational energy is called a _phonon,_ by analogy with a photon which is a quantized unit of electromagnetic energy.

3.   ELECTRONIC EXCITATIONS

The absorption spectrum of a typical alkali halide is shown in Figure 9. Photoconductivity measurements show that the first absorption peaks do not correspond to band-to-band transitions which would result in the production of free electrons and holes. Therefore there exist bound states within the forbidden gap between the valence band

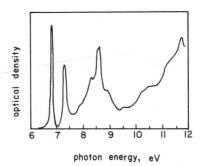

photon energy, eV

Figure 9.   Ultraviolet absorption spectrum of KBr. The first doublet near 7 eV corresponds to an exciton transition from the spin-orbit split halide valence band to an electron orbital of s-like symmetry. The next group of overlapping absorption bands comprises transitions to higher exciton states of s-like symmetry, transitions in which the electron orbital has d-like symmetry, and band-to-band transitions which result in photoconductivity.

and the conduction band. These bound states are called exciton levels and correspond to the formation of a hole-electron pair which migrates as an entity - the <u>exciton</u> - through the crystal.

The absorption spectrum of the silver halides shows a long tail which is due to indirect optical transitions in which a phonon is emitted or absorbed simultaneously with the absorption of a photon.

Interest in the organic solid state is of comparatively recent origin but there is now considerable information available about some of the simpler organic crystals. Anthracene has been studied in particular detail and will therefore be used as an example. The absorption spectrum of anthracene measured with polarized light is shown in Figure 10. Curve <u>a</u> results when the electric vector of the incident radiation lies in the ac plane and curve <u>b</u> corresponds to the electric vector polarized parallel to the b axis of the crystal. The intense absorption bands between

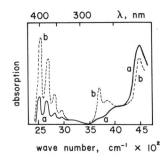

Figure 10.   Absorption spectrum of anthracene. The curve marked <u>a</u> was measured with the electric vector of the polarized light in the ac plane. The curve <u>b</u> was measured with the electric vector parallel to the <u>b</u> axis. [After A.V. Bree and L.E. Lyons, <u>J. Chem. Soc.</u>, 5206 (1960)].

3500 and 4000 Å are observed in solution as well as in the crystalline state but both crystal spectra are shifted to larger wavelengths and the bands of the b spectrum are more intense than those of the a spectrum. The orientation of the molecules in the crystal is such that light polarized parallel to the b axis has its electric vector more nearly parallel to the short axis of the molecule. Thus the b spectrum shows a higher absorption coefficient for these transitions than does the a spectrum.

The most intense absorption in solution is the band at 2500 Å which is associated with transitions polarized parallel to the long axis of the molecule. In the crystal this band is split into two components: in the one at 2700 Å the b spectrum is more intense but in the component at about 1900 Å the a spectrum is more intense. This splitting of the exciton absorption in anthracene was first explained by Davydov and thus bears his name. It arises essentially from the fact that when a single molecule is excited by absorbing a photon the state function for the crystal must be properly antisymmetrized linear combination of state functions for the N-1 molecules in their ground states and the single excited molecule. This leads to a splitting of the single energy state that exists for the isolated molecule. Detailed calculations have been performed by Craig and Walmsley[4]. Their results show that for long axis transitions the splitting is substantial and is such that there is a decrease in the transition energy for transitions with the electric vector polarized along the b axis of the crystal and an energy increase for transitions with the electric vector polarized in the ac plane. Thus we expect a blue shift for the a spectrum and a red shift for the b spectrum, as found experimentally. The Davydov splitting for the short-axis transitions is too small to be revealed in the spectrum but a red shift is predicted again in accordance with experimental results.

---

4.   D.P. Craig and S.H. Walmsley, "Excitons in Molecular Crystals", W.A. Benjamin, New York, 1968.

Anthracene displays a blue fluorescence which is associated with the decay of singlet excitons formed by the $^1A_g \rightarrow {}^1B_{2u}$ transition. The singlet lifetime is 2×10^{-8} sec. There is also a long-lived blue fluorescence with a lifetime of $2 \rightarrow 20 \times 10^{-3}$ sec depending on the purity of the crystal. This is due to a radiationless transfer to the lowest triplet state $^3B_{2u}$, followed by a recombination of two triplet excitons to yield a singlet exciton, followed by a transition to the ground state with emission of the blue fluorescence. The singlet-triplet $^1A_g \rightarrow {}^3B_{2u}$ absorption can be observed using a ruby laser; no phosphorescence from the $^3B_{2u}$ is observed in anthracene crystals although it has been observed from solid solutions of anthracene in benzene.

Whereas pure anthracene irradiated with light of 3400 Å displays the blue fluorescence associated with the $^1B_{2u} \rightarrow {}^1A_g$ transition, anthracene doped with 1 part in 10^4 of tetracene shows the green fluorescence characteristic of pure tetracene. This suggests efficient energy transfer over long distances. The mechanism is not that of resonance transfer but of exciton migration. Excitons are trapped by the tetracene molecules and recombination gives the green fluorescence. In pure anthracene, the excitons will be trapped by imperfections which are probably dislocations.

4. SOME REPRESENTATIVE PHOTOCHEMICAL REACTIONS IN THE SOLID STATE

The Photographic Process

A photographic negative is an emulsion consisting of tiny microcrystals of AgBr (or AgBr + AgI) suspended in gelatine. These microcrystals are thin triangular or hexagonal platelets whose thickness is about one tenth of the linear dimension of their flat faces which varies from several hundredths of a μm up to about 10 μm. The large

226

faces are parallel to $\{111\}$ planes[5]. The most sensitive grains of a high speed emulsion can be rendered developable by the absorption of only three or four quanta. The initial exposure forms a small amount of silver called the 'latent image'. Development increases the number of silver atoms formed in the latent image by a factor of around 10^9. Prolonged exposure leads to the formation of visible silver without development; this is called the 'print out effect'. Sensitization of the film is accomplished by controlled addition of impurities such as sulfide and gold ions. Such impurities undoubtedly are adsorbed on the surface of the silver halide microcrystals and may be incorporated in their bulk as well.

Early ideas on the mechanism of the formation of silver involved extrapolation of knowledge about large single crystals of AgCl and AgBr. It was known quite early on that irradiation with light produced mobile photoelectrons and that interstitial Ag_i^+ were quite mobile. The Gurney-Mott theory[6] supposed that holes were not mobile, that the photoelectrons diffused to trapping sites or 'sensitivity specks' either on the surface or within the grain and were then neutralized by migration of an interstitial Ag_i^+ to the negatively charged trap. Although explaining a number of

---

5. The standard notation used to designate crystal planes is by means of their Miller indices. The Miller indices are found by calculating the reciprocals of the intercepts along the a_1, a_2, a_3 axes, these intercepts being measured in units of the appropriate primitive translational vector. The Miller indices are the smallest set of integers that result after removal of fractions from these reciprocals. They are enclosed in parentheses to denote a particular crystal plane or in braces to denote a set of equivalent crystal planes.

6. R.W. Gurney and N.F. Mott, "Electronic Processes in Ionic Crystals", Oxford University Press, 1938.

the features of the formation of the latent image, the theory is deficient in two respects. Firstly holes are more mobile than Ag_i^+ and so ought to recombine preferentially with the trapped electrons. Secondly, at room temperature there are very few Ag_i^+ present because of the presence of M^{2+} impurity ions accidentally present in the silver halide. The theory was therefore modified by Mitchell and Mott[7]. They considered that holes would be trapped at the surface of the grain thus releasing an interstitial Ag_i^+ which diffused through the grain until it combines with an electron at a trap. At the early stages of formation of a silver nucleus it is difficult to say whether the electron is trapped first and attracts Ag_i^+ or the Ag_i^+ is trapped first and attracts the electron.

An important series of experiments were performed at Bristol by Mitchell and his collaborators. For example, Hedges and Mitchell[8] showed that in thin sheets of AgBr, print out silver decorates the dislocation network. Later experiments by Hamilton and Brady[9] showed that if an emulsion was exposed to light in an electric field and then developed, silver specks were formed predominantly on the surface of the grain on its side towards the anode. This seems to indicate that electrons migrated in the field to surface traps and that Ag_i^+ were then attracted by the Coulomb interaction arising from the net negative charge of the electron and trap. Since the electrons migrate in the applied field the traps in this experiment can be positively charged, negatively charged or neutral but positively charged jogs in dislocation lines, which carry a charge of magnitude $\frac{1}{2}e$, would fit the results of this experiment as

7.   J.W. Mitchell and N.F. Mott, <u>Phil. Mag.</u>, 2, 1149 (1957).

8.   J.M. Hedges and J.W. Mitchell, <u>Phil. Mag.</u>, 44, 223 (1953).

9.   J.F. Hamilton and L.E. Brady, <u>J. Appl. Phys.</u>, 30, 1893, 1902 (1959).

well as the observations of Hedges and Mitchell[8].

X-ray microbeam techniques have shown that the tabular microcrystals of silver halide often contain one, more usually two, twin planes parallel to the $\{111\}$ faces. Such stacking fault twins arise when the stacking sequence is altered from ABCABCAB to the mirror-image sequence ABCACBAC and could be caused if a $\{111\}$ layer were nucleated with an orientation rotated by $60°$ with respect to the layer below. Electron micrographs of replicas of heavily exposed grains show triangular protrusions on the $\{111\}$ surfaces following heavy exposure and the consequent formation of print out silver. Such protrusions are indicative of dislocation loops moving out in $\langle 110 \rangle$ directions from silver particles deposited at the twin planes. These experiments, and those of Hedges and Mitchell, highlight the important role of crystal imperfections in chemical reactivity.

Photolysis of azides

We now consider briefly a photochemical reaction, the decomposition of azides

$$2MN_3 \rightarrow 2M + 3N_2$$

for which there is much less known about the physical properties of the crystals. Here M denotes an alkali metal, an alkaline earth metal, or a heavy metal such as Pb, Ag or Tl. In interpreting the experimental facts we must argue partly by analogy. The general forms of the rate curves are shown in Figure 11. The initial rate of decomposition is proportional to I^2 where I is the intensity of ultraviolet light of wavelength 2537 Å. Radiation of this wavelength produces localized excitons which will be trapped, probably at anion vacancies to form short-lived 'complexes'. Trapping of a second exciton would lead to the formation of nitrogen gas and two F centers. Repetition of the process would result in the formation of F center aggregates, F_n, which recrystallise to form metallic nuclei. This mechanism is represented by equations

$$N_3 + h\nu \rightarrow [N_3]^* \tag{17}$$

$$[N_3^-]^* + \boxed{-} \rightarrow N_3 \boxed{-} e \tag{18}$$

$$[N_3^-]^* + N_3 \boxed{-} e \rightarrow 3N_2 + \boxed{-}\boxed{-}\boxed{-} \overset{e\ e}{} \tag{19}$$

$$F_n \rightarrow M_n \tag{20}$$

If the original trap (depicted above as, but not necessarily, an anion vacancy) becomes surrounded by metal it becomes ineffective, thus explaining the deceleratory nature of the reaction. Fe^{3+} ions, and probably dislocations, can also

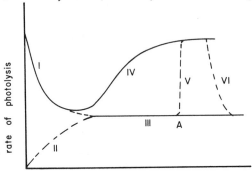

number of azide ions decomposed

per unit area

Figure 11.　　Rate of photolysis of azides as a function of the amount of decomposition [Schematic, after P.W.M. Jacobs, J.G. Sheppard and F.C. Tompkins, 5th International Symposium on Reactivity of Solids, p. 509 (Elsevier, Amsterdam, 1965); P.W.M. Jacobs and A.R.T. Kureishy, Can. J. Chem., 44, 703 (1966). Curves I, III, NaN_3, first irradiation. Curves II, III NaN_3 subsequent irradiations, after exposure to N_2 gas. Curves I, IV, KN_3, BaN_6, SrN_6, 1849 Å + 2537 Å, first irradiation. Curves I, III, KN_3, 2537 , first irradiation (water filter). Curve V is obtained with KN_3 if the water filter is removed at A. Curve VI is obtained with KN_3 if the water filter is inserted at B.

function as traps.

The steady-state process III (see Figure 11) in sodium azide requires the presence of metallic nuclei while the acceleratory process IV (Figure 11) in potassium azide requires the presence of 1849 Å radiation. Quanta of this wavelength will produce delocali ed excitons but they should also do so in sodium azide and the difference in kinetic behaviour between these two similar salts has not yet been explained completely satisfactorily. Equation (19) is one route for the decomposition of two trapped excitons. Other possibilities (with T denoting the trap) are

$$[N_3^-]^* \ + \ N_3Te \ \rightarrow \ N_4^- \ + \ \boxed{-} \ eT \ + \ N_2 \qquad (21)$$

and

$$[N_3^-]^* \ + \ N_3Te \ \rightarrow \ N_2^- \ + \ \boxed{-} \ eT \ + \ 2N_2 \qquad (22)$$

Both N_4^- and N_2^- centers have been detected by esr experiments, N_2^- centers being the more unstable. If the traps are anion vacancies in sodium azide and if these routes are preferred to (19) in sodium azide then the primary electron excess center formed is the F_2^+ center. It has been suggested[10] that the reason for the absence of the acceleratory region in NaN_3 might be connected with bleaching of F_2^+ centers by 5461 Å light present in the output of mercury lamps used in this work.

There is also a post-irradiation dark reaction in sodium azide and it has been suggested that this is connected with the bimolecular decomposition of N_4^- centers[11].

10. D.A. Young, "Progress in Solid State Chemistry", in press. This reference contains a good critical review of the esr and optical information about color centers in azides.

11. P.W.M. Jacobs and A.R.T. Kureishy, J. Chem. Soc., London, 4723 (1964). For the latest work on N_4^- and N_2^- in KN_3 see L.D. Bogan, R.H. Bartram and F.J. Owens, Phys. Rev. B6, 3090 (1972).

$$N_4^- \quad \rightarrow \quad N_3^- \quad + \quad N \qquad\qquad (23)$$

$$N + N \quad \rightarrow \quad N_2$$

Dimerisation of cinnamic acids and of anthracene

The dimerisation of the trans-cinnamic acids is controlled by the geometry of the lattice[12]. The trans-cinnamic acids crystallize in three types of structures (see Figures 12a, b and c). In one of these, designated α, the double bonds in adjacent molecules are < 4 Å apart and are related by a crystallographic center of symmetry. These are converted photochemically into centric dimers (α-truxillic acids). In another type (β) the molecules containing the nearest double bonds are again < 4 Å apart. The β-type acids are converted into dimers of mirror symmetry (β truxinic acids). In the third type (γ) the nearest double bonds are 4.7 - 5.1 Å apart and these acids are photochemically stable. It was this kind of evidence that led to the formulation of the theory of topochemical preformation (Cohen and Schmidt, loc. cit.).

The photochemistry of the anthracenes shows, however, that structural features other than the crystallography of the perfect crystal are involved in solid state photochemical dimerizations. In 9-cyanoanthracene (and in one form of the dimorphic 9-chloro-) ultraviolet irradiation yields centric (head-to-tail or trans) dimers whereas mirror symmetric (head-to-head or cis) dimers would be expected from the topochemical preformation theory (type β, Figure 13). In view of the role of crystal defects in the photochemical dimerization of unsubstituted anthracene[13] (type α, Figure 13), it seemed likely that the explanation of the

---

12. M.D. Cohen and G.M.J. Schmidt, J. Chem. Soc., London, 1996 (1964), and following parts of this series.

13. J.M. Thomas and J.O. Williams, Chem. Comm., 432 (1967).

Figure 12.   The three types of crystal structure
              occurring in the <u>trans</u>-cinnamic acids.

(a)   The packing arrangement in a type-α structure, α-
      <u>trans</u>cinnamic acid, seen along [001]. The nearest
      double bonds are 3.56 Å apart and adjacent molecules
      are clearly antiparallel. Dimerization results in a
      head-to-tail dimer of molecular symmetry $\overline{1}$ = S_2 = i.

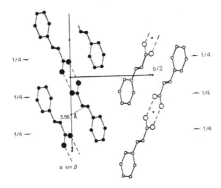

(b)   The packing arrangement in p-chloro-<u>trans</u>-cinnamic acid
      which is typical of type-β structures, seen along [010].
      The shortest distance between C = C double bonds is
      3.87 Å and occurs parallel to the b axis. Nearest
      neighbor molecules are parallel to one another so that
      the dimers have molecular symmetry m.

233

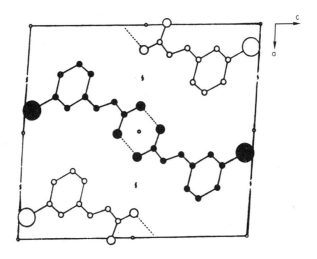

(c)   The packing arrangement in m-chloro-<u>trans</u>-cinnamic
      acid which is an example of the photochemically stable
      γ-type structure.   The shortest distance between
        C = C   double bonds is 4.93 Å and this exceeds the
      critical distance above which dimerisation does not
      occur.

perverse behavior of the 9-substituted compounds also lay
in a mechanism involving crystal imperfections.   This has
recently been shown to be so[14].   There is a strong align-
ment of dimer nuclei along the [001] direction on (100)
planes.   The nuclei frequently displayed preferred directions of growth namely [021], [02$\bar{1}$], [01$\bar{5}$], [0 1 10] and
[0 1 1$\bar{0}$].   This suggests a connection between nucleation
and dislocations.   There are a number of reasons why dislocations might function as preferred sites for photochem-

---

14.   M.D. Cohen, Z. Ludmer, J.M. Thomas and J.O. Williams
        Chem. Comm., 1172 (1969).

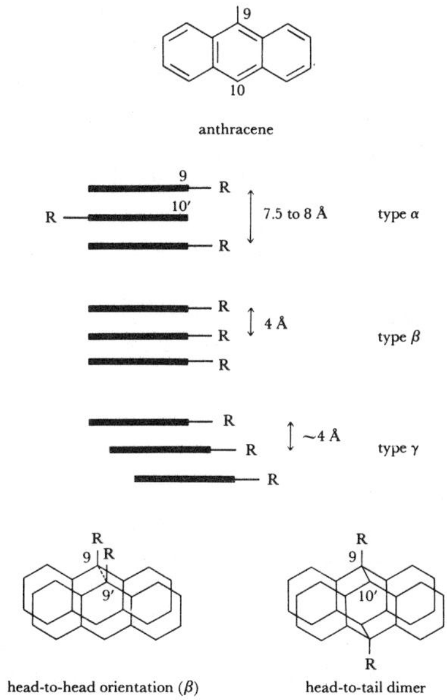

anthracene

type α

7.5 to 8 Å

type β

4 Å

type γ

~4 Å

head-to-head orientation (β)

head-to-tail dimer

Figure 13.   A schematic representation of the crystal
structure of 9-substituted anthracenes. In
type α, for example one each of the polymorphs of 9-methyl- and 9-amidoanthracene,
nearest neighbor molecules have a head-to-
tail orientation and centric dimers can form
without steric hindrance. In type β adjacent molecules are translationally equivalent
and the formation of centric dimers linked
in the 9-10' positions requires, in the
perfect structure, a rotation of one of the
molecules. In type γ, R = H, the molecules
are not strongly overlapped and one expects
anthracene to be photochemically inert in
the perfect crystal.

235

chemical reaction.

(i)   If the product phase possesses a molecular volume different from that of the reactant, then the first-formed molecules of the new phase can be accommodated in the regions of dilation or compression with a minimum of strain.

(ii)   Since the normal lattice symmetry is disrupted at the dislocation, molecules at dislocations will have different energy levels to those in perfect crystal. Thus electronic excitation may be easier at these sites or they may act as traps for mobile excitons.

(iii)   Molecular rotation, if necessary to form the product, will be facilitated in the dilated regions of edge dislocation cores.

Cohen et al.[15] have stressed the possible importance of stacking faults in the photodimerization of 9-cyanoanthracene. The dissociation of unit strength dislocations on the (221) plane into two partial dislocations produces stacking faults in which the nearest-neighbor orientation, which is nearly <u>cis</u> in the perfect crystal, is now such that half the total number of molecules in the stacking fault ribbon are in a <u>trans</u> configuration. The <u>trans</u> dimer is thus nearly pre-formed in such regions, which as we have seen in (ii) above, are favorable sites for the initiation of photochemical reactions.

---

15.   M.D. Cohen, Z. Ludmer, J.M. Thomas and J.O. Williams, <u>Proc. Roy. Soc. London,</u> A 324, 459 (1971)

GENERAL REFERENCES

[1]   W.J. Moore, "Seven Solid States", W.A. Benjamin, New York, 1967.

[2]   N.B. Hannay, "Solid State Chemistry", Prentice Hall, Englewood Cliffs, New Jersey, 1967.

[3]   A.K. Galwey, "Chemistry of Solids", Chapman and Hall, London, 1967.

[4]   F.A. Kroger, "The Chemistry of Imperfect Crystals", Interscience, New York, 1964.

[5]   D.A. Young, "Decomposition of Solids", Pergamon, Oxford, 1966.

CHAPTER 10

PHOTOCHROMISM

Introduction

When rocks containing the mineral hackmanite are broken the freshly exposed surface is a deep magenta in color. This color fades rapidly in sunlight but may be reproduced in natural or synthetic sodalites by exposure to ultraviolet or X-irradiation. This is an example of photochromism or the phenomenon of a reversible change in color brought about by exposure to electromagnetic radiation. Photochromism is to be distinguished from luminescence in which the absorption of radiation raises the absorbing centers to an unstable excited state A* from which they revert to the ground state by emission of radiation, usually of different frequency. In photochromic substances absorption of radiation leads to a quantum mechanically stable but thermodynamically metastable state B from which it may revert to the original state A by absorption either of light of a different wavelength or of thermal energy. Three conditions are thus necessary for a material to be photochromic; the state B should be thermally stable[1], it should absorb radiation in a different region of the spectrum to A, and the radiation or thermally induced change B → A should occur. Despite these rather restrictive

---

1. A qualitative term - the lifetime of B in the absence of radiation will depend on the thermal activation energy for the return of B to A.

conditions a wide range of organic and inorganic materials
have been found to be photochromic either in the solid or
glassy states or in solution. In some substances a change
in color can be brought about by heat; this is the related
phenomenon of thermochromism.

The general photochromic change is illustrated schematically by

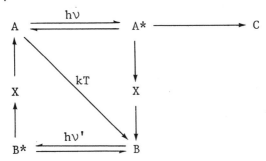

The excited state A* may form the photochromic state
through one or more intermediate states designated X or it
may undergo an irreversible chemical change to produce C.
While chemical changes are not excluded from the photochromic scheme- and indeed one of the best known and techno-
logically applied examples of photochromism involves a reversible chemical change - irreversible photochemistry is
seen to be an undesirable feature resulting in a reduction
of the efficiency of the reversible change A $\rightleftarrows$ B and constituting a potential source of fatigue in a photochromic
device.

The remainder of this chapter will consist of a systematic consideration of the mechanisms of photochromic
changes in both organic and inorganic materials. It will
be seen that, mechanistically speaking, organic chemical
systems are the more versatile and that while nearly all
the inorganic photochromic systems, including most of those
that involve a chemical change, comprise an electron transfer process, photochromic changes in organic chemical systems may involve molecular rearrangement, homolytic or
heterolytic cleavage, or redox changes. No attempt will be

made to catalogue the extremely large number of materials that have been shown to display photochromism. Not only would such a catalogue be a positive handicap in our attempt to understand how photochromic changes occur, but it would be unnecessary because of the reviews of this subject which have already appeared in the literature and which are listed at the end of this chapter.

ORGANIC PHOTOCHROMIC SYSTEMS

Molecular rearrangement

This type of photochromic change includes cis-trans isomerizations and hydrogen transfer processes followed by a geometric rearrangement of the molecule. Typical and striking examples of photochromic cis-trans isomerization are provided by the thioindigo dyes[2]. The two isomers are in equilibrium in solution, the trans form (I) being the

I   II

more stable at ordinary temperatures. Absorption at long wavelength shifts the equilibrium in favor of the cis form and the change may be reversed by irradiating in the ultraviolet or short wavelength region (blue) of the visible spectrum. Irradiation produces an excited singlet which may undergo intersystem crossing to the triplet with subsequent isomerization. The most stable configuration for the lowest triplet is the $90°$ twisted configuration from which

2.   G.M. Wyman and W.R. Brode, J. Amer. Chem. Soc., 73, 1487 (1951).

the molecule may relax into either the trans or cis forms.
Similar photochromic changes occur in the 4-amino- and 4-
hydroxy-azo dyes but thermal equilibration is faster and it
is consequently more difficult to determine absorption
spectra for the individual cis and trans forms.

A beautiful example of hydrogen-transfer isomerization
is provided by the anils[3] of salicylaldehyde[4,5]. With the
exception of anils derived from 3-hydroxy 2 naphthaldehyde,
all anils with a hydroxyl group ortho to the C = N- group
show photochromism in rigid glasses. Solutions of 1,2- and
2,1-hydroxy-naphthaldehyde anils, at sufficiently low temperatures, develop a thermochromic absorption band which is
similar to the spectrum of the photochromic anils. The
importance of an ortho hydroxy group is confirmed by the
absence of photochromy when benzaldehyde or p-hydroxy-
benzaldehyde is condensed with aniline, or the ortho
hydroxy group is methylated. Many of the ring substituted
anils occur in two polymorphic forms, one of which is
photochromic and the other thermochromic. These observations can be interpreted by postulating a two-step molecular rearrangement. III is the stable anil which is in
tautomeric equilibrium with IV. Thermochromism involves a
shift in this equilibrium towards IV. Photochromism accompanies the cis-trans isomerization of IV into V through
rotation about the C = C bond. The thermal activation
energy for the reverse (color-fading) reaction is that required for the thermal trans-cis isomerization. The long-
wavelength absorption at 540-580 nm in the photochromic

3.   The condensation products of aldehydes and primary
        amines with the general formula R-CH = N-R' are
        known as anils or Schiff bases.

4.   G.M.J. Schmidt, Proceedings of the Thirteenth Con-
        ference on Chemistry, Brussels, 1965 (Interscience,
        London, 1967).

5.   M.D. Cohen and G.M.J. Schmidt, Proceedings of the
        Fourth International Symposium on the Reactivity of
        Solids, Amsterdam, 1960 (Elsevier, Amsterdam, 1961).

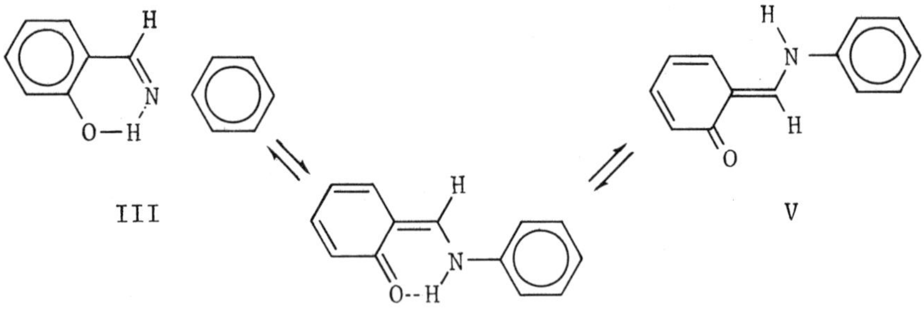

colored species is assigned to the n-π* transition in the
quinoid isomer V where the oxygen atom is no longer involved in hydrogen bonding. The reason for the different behavior of the two polymorphs lies in their crystal structures. In the photochromic form the molecules are non-
planar and there are no close intermolecular contacts normal to the ring. Consequently the activation energy $\Delta E'$
for the cis-trans isomerization step is not much higher
than in the free molecule. In the thermochromic form the
molecules are planar with intermolecular spacing normal to
the rings of only 3.3 Å. This close contact increases $\Delta E'$
and also stabilizes the quinoidal form IV through intermolecular (dipole-dipole or hydrogen bonding) interactions
(see Figure 1).

Dissociation

Photochromic changes which depend on heterolytic cleavage occur in the spirans, triarylmethane dyes, polymethine
dyes, indenone oxides, and nitrones. In a comprehensive
review Bertelson[6] [4] distinguishes sixty spiran ring systems some of which are represented by scores of examples.

---

6. General references at the end of this Chapter are
   indicated by [ ].

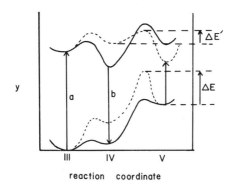

reaction   coordinate

Figure 1.   Energy of the isomers III, IV, V in the
ground and excited states. The continuous
line represents the thermochromic polymorph
and the dotted line indicates the photochromic crystal. Fluorescence b in the thermochromic crystal is appreciably Stokes-shifted
with respect to absorption a in the -OH form
III due to H-transfer in the excited state.
ΔE is the thermal activation energy for the
trans-cis isomerization (photochromic fading).
$\Delta E'$ is the thermal activation energy for the
cis-trans isomerization in the excited state
and is > 0 since a low temperature limit
exists below which photocoloration of the
solid no longer occurs.

The photochromism of literally hundreds of spirans has been
investigated sometimes in great detail. The spirans (spiropyrans) contain a 2H pyran ring with carbon atom 2 involved
in a spiro linkage. A specific example is 1,3,3-trimethyl-
indolino-benzopyran VI which is colorless but may be trans-
formed by absorption of light into the colored form VII.
This involves a substantial change in molecular geometry
and yet at the instant of C-O bond rupture the molecule
must still possess the geometry of the original spiran
molecule. This state X must correspond very closely to the
transition state for the thermal fading reaction in which

243

VI   VII

the molecule assumes just the right geometry prior to ring closure. The molecule X possesses an excess of vibrational energy and as vibrational deactivation proceeds the geometry gradually approaches that of the metastable colored species B (of which there may be more than one form).

Bertelson [4] believes that this vibrational deactivation proceeds through a series of isomers with the possibility of forming the colored species from one or more of these isomers. Failure to do so results in reformation of the uncolored species A.

A different mechanism has been proposed by Becker[7] based on the discovery that formation of the colored species may require vibrational energy in the first excited state (Figure 2). In this mechanism there exists a competition between internal conversion within the S_1 manifold

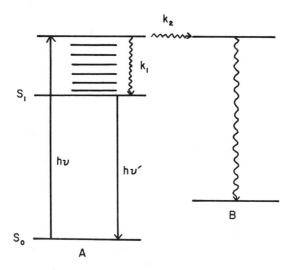

Figure 2.   Energy level diagram

(k_1) and transformation to an excited state of the colored species, B* (k_2), which then undergoes radiationless deactivation to the ground state, B. At low temperatures the molecules taking the internal conversion route revert to the ground state A by fluorescence, emitting a photon of energy $h\nu'$. The chief difference to the isomerization mechanism is that the latter provides for non-radiative deactivation to the ground state via a sequence of geometric isomers and provides for formation of the colored species from one or more vibrationally-excited states of these isomers.

---

7.   R.S. Becker, E. Dolan and D.E. Balke, J. Chem. Phys., 50, 239 (1969).

The open (B) and closed (A) forms are analogous to the trans and cis isomers about the 3-4 double bond. Because of lower bond order in the excited state rotation of the open form into the closed form will occur more readily in the excited state. The thermal activation energy for the photoerasure (photobleaching) reaction could then be associated with the potential barrier for rotation in the excited state[8] although a competing view[9] is that a thermal isomerization step in the ground state of B occurs after the photochemical formation of a cis isomer X.

Colored solutions of many triarylmethane dyes are decolorized when treated with CN^-, OH^- or HSO_3^-, forming the dye leucocyanide, leucohydroxide or leucobisulfite. These colorless solutions containing the dye leuco derivatives may be colored by exposure to ultraviolet light. Photochromic behavior may be observed not only in dilute solution but also in gels, plasticized resins and in thin films. A comprehensive list is given by Bertelson [4]. While it is clear that in solvents of high dielectric constant there is heterolytic cleavage of the nitrile group of the leucocyanide to yield the colored dye, details of the mechanism and that for the thermal fading reaction remain obscure. A frankly speculative modification of a mechanism due to Brown et al.[10] is

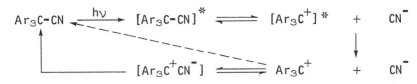

8.   O. Bloch-Chaudé and J.-L. Masse, Bull. Soc. Chim. France, 625 (1955).

9.   R. Heligman-Rim, Y. Hirshberg, and E. Fischer, J. Phys. Chem., 66, 2465 (1962).

10.   G.H. Brown, S.R. Adisesh and J.E. Taylor, J. Phys. Chem., 66, 2426 (1962).

Bertelson [4] includes a further (unspecified) colorless intermediate "X" in equilibrium with the dissociated ions and the ion pair. The crucial steps would appear to be the formation and removal of the solvation sheath around the dye cation. A feature of this system which, if not eliminated, would be a handicap to possible commercial exploitation is the fatigue which results from photodegradation of the dye and also from irreversible processes involving the anion[11]. Of the many classes of substances that have been added to photochromic dye systems in experiments designed to reduce fatigue, the most promising appears to be enzymes.

Tetrachloro-1 (4H) naphthalenone VIII displays photochromism both at room temperature in carbon tetrachloride solution and in glasses at low temperatures, this behavior being typical of a number of substituted quinols.

VIII   IX

VIII is polymorphic, only the β isomer being photochromic. A similar situation exists for the 2H compound but a detailed explanation of these topochemical factors is lacking. Chemical evidence for the existence of free radicals led to the proposal[12] that the photochromic change comprises the

---

11.   R.N. Macnair, Photochem. Photobiol., 6, 779 (1967).

12.   G. Scheibe and F. Feichtmayr, J. Phys. Chem., 66, 2449 (1962).

photodissociation of a Cl atom from the 4 position to give the naphthoxy radical IX. However, the absence of an esr signal[13] from the irradiated β-form of VIII goes against this proposal.

Redox systems

A photochromic redox system consists of donor and acceptor components each of which must exist in at least two stable oxidation states. One of these must obviously be colored and the other uncolored (or of a different color). The primary act of absorbing a photon must result in sufficient charge relocalization to make the subsequent ionization step reasonably probable and this is especially likely in charge transfer transitions. Several phenothiazine dyes function as photochromic redox systems. The photochemical bleaching of thionine will serve as an example. The stoichiometry of the equilibrium between thionine (Lauth's violet) and leucothionine is expressed by the equation

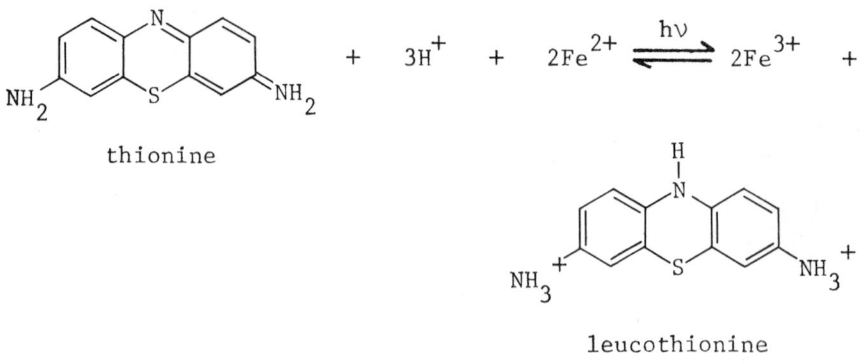

$$+ \quad 3H^+ \quad + \quad 2Fe^{2+} \quad \underset{}{\overset{h\nu}{\rightleftarrows}} \quad 2Fe^{3+} \quad +$$

thionine

leucothionine

---

13.   H.S. Gutowsky, R.L. Rutledge, and I.M. Hunsberger, J. Chem. Phys., 29, 1183 (1958).

The solution displays fluorescence associated with the return of the excited singlet thionine to its ground state. Intersystem crossing yields the excited triplet which accepts two protons from solution and also an electron from the ferrous ion donor. The resulting semithionine then disproportionates into thionine and the colorless leucothionine. Denoting thionine by Th^+ this kinetic mechanism is summarized by the equations:

$$Th^+(S_o) \xrightarrow{h\nu} Th^+(S_1)$$

$$Th^+(S_1) \xrightarrow{h\nu'} Th+ (T_1)$$

$$Th^+(T_1) + 2H^+ + Fe^{2+} \rightarrow [ThH_2]^{2+} + Fe^{3+}$$

$$2[ThH_2]^{2+} \rightarrow Th^+ + [ThH_3]^{3+} + H^+$$

This system is extremely stable and appears to be completely reversible in the absence of oxygen.

Chlorophyll also undergoes reversible photobleaching either in oxygen-free solution in appropriate solvents or in glasses at low temperatures. Although the exact identification of intermediates is yet to be accomplished it seems clear that an electron is transferred from an excited state of the chlorophyll molecule (C) to a solvent molecule or some other added acceptor such as quinone (Q). In the latter case

$$C + h\nu \rightarrow C*$$

$$C* + Q \rightarrow C^+ + Q^-$$

Esr studies[14] have revealed a slow bimolecular decay of semiquinone

$$Q^- + Q^- \rightarrow Q^= + Q$$

---

14. B.J. Hales and J.R. Bolton, _J. Amer. Chem. Soc._,

INORGANIC SYSTEMS

Alkali halides

If a potassium chloride crystal containing F centers
(Chapter 10) is irradiated at low temperatures, the F band
bleaches and simultaneously a much broader absorption occurs
on the long wavelength side of the F band. This is called
the F' band. Irradiation within the F' band regenerates
the F band so that this system represents a simple example
of inorganic solid state photochromism. The F' band is
thermally unstable and decomposes on warming, the F band
being regenerated. Irradiation in the F band at low temperatures produces photoconductivity and the temperature
dependence of photoconductivity patterns that for F $\rightarrow$ F'
conversion, both showing a steep fall-off below 150 K.
Corresponding with the fall-off in photoconductivity there
is an increase in F-center fluorescence.

A characteristic quantity in photoconductivity is the
mean range w of the photoelectrons. The probability p of
capture of a photoelectron is proportional to the trap
density N and to the velocity of the electrons v (= u$\mathcal{E}$
where u is the mobility and $\mathcal{E}$ the electric field). The
proportionality constant is called the cross section σ so
that

$$p = \sigma Nv \qquad (1)$$

Hence the mean time that the electrons are travelling
through the crystal in the conduction band is

$$\tau = \frac{1}{p} = \frac{1}{\sigma Nv} \qquad (2)$$

Their mean range is then

$$w = v\tau = \frac{1}{\sigma N} \qquad (3)$$

Experimentally, the mean range turns out to be inversely
proportional to the number of F centers showing that the F
centers themselves are the traps. Hence the mechanism for
this photochromic change is

$$F \xrightleftharpoons[h\nu'']{h\nu} F^* \tag{4}$$

$$F^* \xrightleftharpoons[]{kT} e + \boxed{-} \tag{5}$$

$$e + F \xrightleftharpoons[h\nu',kT]{} F' \tag{6}$$

$\boxed{-}$ denotes an anion vacancy. At temperatures above 150 K in KCl thermal ionization of electrons from the excited state F* occurs in preference to fluorescence (hν''). At higher temperatures still the anion vacancies compete succ- essfully with the F centers as traps owing to the thermal instability of the F' center (see Figure 3). On the basis

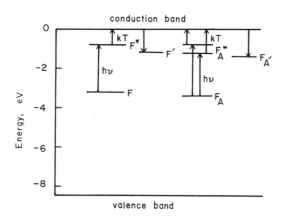

Figure 3.   Energy level diagram for KCl showing the stability of the F, F', F_A and F_A' centers.

of this model a maximum quantum yield of 2 is predicted for the F to F' conversion, in either direction. This is con- firmed by experimental measurements. The fall-off in quantum yield for F' → F conversion and the increase in that for F → F' conversion in the region 100-150 K are both due to the improved efficiency of the thermal ionization of F centers which occurs in this temperature region, Equation (5).

Confirmation of steps (4) and (5) comes from measurements of the fluorescence lifetime as a function of temperature. The mean lifetime τ^* of an electron in the excited state F* is given by

$$\frac{1}{\tau^*} = \frac{1}{\tau_r} + \frac{1}{\tau_i} + \frac{1}{\tau_q} \tag{7}$$

where τ_r is the radiative lifetime, $1/\tau_i$ is the probability per unit time for thermal ionization,

$$\frac{1}{\tau_i} = \frac{1}{\tau_o} \exp(-\Delta E /kT) \tag{8}$$

τ_o being a frequency factor of the order of a lattice vibrational frequency, and $1/\tau_q$ is the probability for nonradiative decay to the ground state. This last process is improbable at low temperatures so that

$$\tau^* = \frac{\tau_r}{1 + (\tau_r/\tau_o)\exp(-\Delta E /kT)} \tag{9}$$

and the relative fluorescence yield is

$$\varphi_r = \frac{1/\tau_r}{1/\tau^*} = \frac{1}{1 + (\tau_r/\tau_o)\exp(-\Delta E /kT)} \tag{10}$$

The temperature dependence of the fluorescence yield is well-fitted by Equation (10) with $\Delta E \simeq 0.15$ eV and τ_r of the order of 10^{-6} sec. Such a long lifetime is unexpected, in view of the high oscillator strength in the F band, and is explained by relaxation which occurs in the excited state resulting in the F center electron being very spread out over the neighboring ions.

This photochromic system has been described in some detail because of its simplicity and because of the detailed understanding that now exists concerning F centers, rather than for its practical use. The F' centers are rather too unstable for this system to be incorporated in a workable device at ordinary temperatures. This objection can be overcome by selective doping of the alkali halide with a cation of smaller ionic radius e.g. KCl with Na^+ of Li^+. An F center formed by trapping an electron at an anion vacancy adjacent to the foreign cation is called F_A center. The lower symmetry of the F_A center (C_{4v}) compared to the F center (O_h) causes a splitting of the threefold degenerate F* state into two components, a non-degenerate one associated with a transition (F_{A1}) polarized in the direction of the neighboring cation and a twofold degenerate one associated with transitions (F_{A2}) polarized in the plane perpendicular to this direction. Thus the single F absorption band is replaced in the doped crystals by two overlapping bands which exhibit dichroism. The substitution of a lighter cation in one of the nearest neighbor positions also has the effect of lowering the energy of the ground state of the F_A' center (see Figure 3) as shown by the shift in the F_A' absorption band to shorter wavelengths. This permits the use of the photochromic $F_A \rightleftarrows F_A'$ conversion at room temperature.

Sodalites

The chemical composition of sodalite is $6(NaAlSiO_4) \cdot 2NaCl$. The structure of sodalite consists of an open aluminosilicate framework with a group of four Na^+ ions tetrahedrally surrounding a Cl^- ion in the center of each cavity of the framework. This material can be made photochromic by firing it in an inert atmosphere or in hydrogen or deuterium. This treatment induces an absorption band in the ultraviolet at 4.7 eV and irradiation with photons of this energy colors the sodalite pink. The color center responsible for this reversible photochromic effect consists of an electron trapped at a vacant anion site. This identification is reasonably certain being based on the appearance of 13 resolved components in the hyperfine splitting of the esr absorption at 9 GHz. The spin g factor is 2.002 and the interaction of the trapped electron with four ^{23}Na nuclei of spin $I = 3/2$ splits the absorption line into ($2 \times 4 \times 3/2$) $+1 = 13$ components with a hyperfine spacing of approximately

31 G. If the Cl is replaced by Br or I the color of the irradiated sodalite changes progressively from pink to blue. Photochromism is enchanced by doping with S, Se or Te and while the photochromic center in sodalites is identified with reasonable certainty the mechanism of its formation under ultraviolet radiation is not understood at the present time.

Calcium fluoride

Calcium fluoride doped with rare earth ions has been investigated intensively in recent years. When lanthanum, gadolinium, cerium or terbium ions are added to CaF_2 and it is treated under reducing conditions the resulting material shows photochromic effects at room temperature. Consider cerium-doped CaF_2 as an example. The reducing treatment results in some thermally stable centers associated with Ce^{3+} ions but these are not simply single Ce^{3+} ions substituting for Ca^{2+} ions on the cation sub-lattice since they exhibit linear dichroism along the $\langle 111 \rangle$ directions. The optical absorption of the center, however, corresponds to internal transitions on the Ce^{3+} ion. The highest occupied state lies about 3 eV below the conduction band and there is also a vacant level about 2 eV below the conduction band which corresponds to Ce^{2+} ions. Irradiation of the cerium-doped CaF_2 with light of wavelength < 450 nm produces free electrons, as shown by photoconductivity, and some of these are trapped in the vacant states to produce Ce^{2+} ions, which may be identified by their absorption spectrum. Irradiation with light of wavelength > 450 nm causes the Ce^{2+} centers to lose electrons to the conduction band, the electrons being re-trapped by the originally ionized centers to regenerate Ce^{3+} ions. The photochromic change is thus a very simple redox cycle involving cerium ions (Figure 4).

Rutile and titanates

Photochromism in TiO_2 and in the alkaline earth titanates is associated with the presence of iron impurity. Fe^{3+} ions enter the TiO_2 structure substitutionally for Ti^{4+}

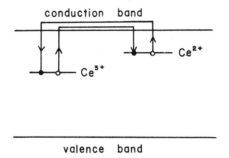

conduction   band

Ce²⁺

Ce³⁺

valence   band

Figure 4.   Electron transfer photochromic change in
cerium-doped calcium fluoride.

ions, each pair of Fe^{3+} ions being compensated for by one $0^=$
vacancy. Some of the Fe^{3+} ions occupy sites adjacent to the
vacancies forming Fe^{3+} $\boxed{-}$ centers. Iron-doped rutile is
colorless or pale yellow and irradiation with light of photon energy > 2.5 eV superimposes a pinkish brown color.
This is due to excitation of electrons from the valence band
into the Fe^{3+} $\boxed{-}$ centers. The reason why electrons are
trapped at these centers rather than at Fe^{3+} ions without
the adjacent vacancy is that the Fe^{2+} ion is larger than can
be accommodated comfortably on a substitutional cation site
in rutile whereas the vacancy permits the larger (albeit
deformed) Fe^{2+} ion to be accommodated. The positive holes
migrate through the valence band until trapped by Fe^{3+} ions
without the associated vacancy (Figure 5a). Thermal or
optical bleaching of the photochromic absorption occurs by
excitation of electrons from the Fe^{2+} $\boxed{-}$ centers to the
conduction band and their re-trapping by Fe^{4+} ions ($\equiv Fe^{3+}$
+ hole) to regenerate substitutional Fe^{3+} ions. The thermal
activation energy for the loss of color is 0.5 eV, much less
than the optical value of ~ 2.5 eV because of the big lattice relaxation that occurs following a (Franck-Condon)
optical transition. The relaxed $[Fe^{2+}$ $\boxed{-}$ $]^*$ state either
overlaps or is just below the conduction band. The coloration and bleaching processes are shown schematically in
Figure 5a, b.

255

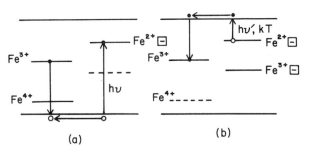

Figure 5.   Coloration (a) and bleaching (b) processes in photochromic iron-doped rutile.

Electron transfer processes between localized centers in non-metallic crystals are thus seen to involve either migration of electrons in the conduction band or of holes in the valence band or a combination of these two processes. Strontium titanate doped with Fe^{3+} ions behaves similarly to rutile and the formation and regeneration of Fe^{3+} can be followed by the changing intensity of the cubic Fe^{3+} esr absorption. The thermal stability of the colored state can be increased by double doping the titanate with iron and molybdenum. Charge compensation now occurs by one Mo^{6+} ion occupying a Ti^{4+} site for every two Fe^{3+} ions that enter the structure substitutionally. In the photochromic process an electron is transferred from Fe^{3+} to Mo^{6+} on irradiation with ultraviolet light. The details of the charge transfer process are not known. The three possibilities are shown in Figure 6. The thermal lifetime of the photochromic state varies markedly with the host crystal orders of magnitude being 10^2, 1 and 10^{-2} seconds for iron-doped calcium, strontium and barium titanate respectively.

Glasses

Electron transfer processes also occur in glassy media. $Na_2O \cdot SiO_2$ transmits light down to 2100 Å. Under X-irradiation an absorption band appears at 5700 Å which is stable but not permanent, decaying over a period of months. If 0.3% Ce^{3+} is added to pure glass, it will now color with ultraviolet light and the coloration decays in 5 minutes instead of months. The Ce^{3+} has induced an absorption band

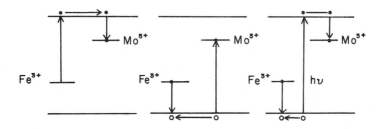

Figure 6.   Possible charge transfer mechanisms in
             strontium titanate doped with iron and
             molybdenum

             (a) electron transfer through the conduction
                  band
             (b) hole transfer through the valence band,
                  and
             (c) electron-hole pair excitation.

in the ultraviolet and irradiation within that band produces Ce^{4+}, the ionized electron being trapped elsewhere in the glass. The type of trapping center has been identified by double doping experiments using Eu^{2+} and Ti^{4+}. Irradiation in the Eu^{2+} band produces the electron transfer reaction

$$Eu^{2+} + Ti^{4+} \rightleftarrows Eu^{3+} + Ti^{3+} \qquad (11)$$

$$4f^6 5d^1 \qquad 3d^0 \qquad 4f^6 5d^0 \qquad 3d^1$$

The electron configurations of the ions are shown. The mechanism of electron transfer is not understood in detail but it presumably involves tunnelling from an excited state of the rare earth ion to a nearby cation with empty d orbitals. The extent to which $O^=$ orbitals may be involved in the tunnelling process is not known at present. The d^1 configuration of Ti^{3+} is unstable in this configuration and the electron transfers back to the Eu^{3+} ion to reform Eu^{2+} in its ground state. Zr^{4+} ($4d^0$) and Hf^{4+} ($5d^0$) behave similarly to Ti^{4+}. The color of the photochromic colored state is due to internal transitions of the ions (Ti^{3+}, Zr^{3+}, Hf^{3+}) with the d^1 configuration.

Silver halide glasses

The first report of a photochromic effect in borosilicate glasses contining silver halides in concentration by volume of 5×10^{-4} and with a particle size of > 50 Å occurred in 1964.[15] The silver halides decompose when irradiated with visible light and the glass darkens; when the light is switched off the transmission of the glass is restored. The photochromic response of the glasses is affected by the rate of cooling which in turn controls the strain to which the AgCl microcrystals are subjected. A small amount of copper is included in almost all silver halide photochromic glasses since it increases the darkening sensitivity by several orders of magnitude. The resistance to fatigue of the silver halide glasses is excellent. One

---

15.   W.H. Armistead and S.D. Stookey, _Science_, 144, 150 (1964).

particular sample[16] has been cycled at a rate of 1 cycle per minute for upwards of 300,000 cycles with no apparent signs of fatigue, the change in optical density per cycle being 0.3.

The overall mechanism is clearly the decomposition of the silver halide (henceforth referred to as silver bromide although AgCl or mixtures of AgCl and AgBr may be used) into particles of metallic silver, which are responsible for the darkening, and bromine which recombine when illumination ceases. The details of the mechanism are not yet resolved although the general outline seems clear enough. Absorption of a photon produces free holes and electrons. Both are mobile. A mobile electron is readily trapped by imperfections which could be stacking faults, cationic jogs along edge components of dislocation lines, or surface defects, to form a silver atom. At this stage the silver atom is vulnerable to attack by a positive hole [Equation (14)] but once three or more silver atoms form by successive trapping of electrons and Ag^+ ions, the silver atom aggregate absorbs Ag^+ ions and while continuing to grow by alternate absorption of Ag^+ ions and trapping of electrons, remains on average positively charged and so repels positive holes. The formation of silver can thus be represented by the following equations:

formation of a free electron and hole by absorption of a photon:

$$Br^- + h\nu \rightarrow e + Br^o \tag{12}$$

trapping of free electron and its neutralization by an interstitial silver ion:

$$e + Ag_i^+ \rightarrow Ag \tag{13}$$

16. G.K. Megla, Appl. Optics, 5. 945 (1966).

17. J.W. Mitchell, J. Phys. Chem., 66. 2359 (1962).

recombination of hole and electron at trap:

$$Ag + Br^o \rightarrow Ag^+ + Br^-$$ (14)

growth of silver nucleus:

$$Ag_n + Ag_i^+ \rightarrow Ag_{n+1}^+$$ (15)

$$Ag_{n+1}^+ + e \rightarrow Ag_{n+1}$$ (16)

The lifetime of positive holes in silver bromide is independent of the mechanical perfection of the crystal indicating that dislocations do not form efficient traps for holes. Foreign ions that can change their valence readily, such as Cu^+ or Fe^{2+}, do, however, trap holes very effeciently and the improvement in photochromic efficiency that results from the addition of copper is undoubtedly due to the efficacy of Cu^+ as a hole trap thus reducing the probability of hole-electron recombination [Equation (14)][18] The Cu^+ ion is regenerated by thermal excitation of an electron from the valence band. Eventually holes are trapped at the surface of the AgBr particle, forming adsorbed bromine atoms and interstitial silver ions (Ag_i^+) the latter diffusing to the silver nuclei to supply the flux of silver ions needed for its growth [Equation (15)]. The fate of the holes is thus represented by the following equations: trapping of hole at cuprous ion:

$$Br^o + Cu^+ \rightarrow Cu^{2+} + Br^-$$ (17)

liberation of hole from Cu^+

$$Cu^{2+} + Br^- \rightarrow Cu^+ + Br^o$$ (18)

trapping of hole at surface, formation of adsorbed bromine atom, and of interstitial Ag_i^+:

$$Br^o + Ag^+ \rightarrow Br(s) + Ag_i^+$$ (19)

---

18.   J. Malinowski. J. Phot. Sci., 16, 57 (1968).

The possible role of cation vacancies in the reformation of holes at Cu^+ traps and their participation in the diffusion of hole-vacancy complexes to the surface, as proposed by Malinowski[19], cannot be assessed completely at present.

GENERAL REFERENCES

[1]   R. Exelby and R. Grinter, Chem. Rev., 65, 247 (1965).

[2]   R. Dessauer and J.P. Paris, "Advances in Photo-chemistry". Vol. 1, p. 275, Wiley, New York, 1963.

[3]   W. Luck and H. Sand, Angew. Chem., 76, 463 (1964).

[4]   "Photochromism", ed. G.H. Brown, Wiley Interscience, New York, 1971.

[5]   G. Gliemeroth and K-H. Madler, Angew. Chem., 82, 421 (1970).

[6]   Z.J. Kiss, Physics Today, 1, 42 (1970).

[7]   G. Jackson, Optica Acta, 16, 1 (1969).

---

19.   J. Malinowski, Contemp. Phys., 8, 285 (1967).

CHAPTER 11
INDUSTRIAL APPLICATIONS
OF PHOTOCHEMISTRY

The suggestion that photochemistry may have something to offer industry is not new. Such a point of view was clearly and unambiguously stated sixty years ago by Ciamician[1] in an address to the chemical industry in the United States. It is the purpose of this Chapter to review progress since that time, and to discuss some of the chemical and economic factors which are likely to influence the future.

That _some_ photochemical processes are important is undeniable since vision and photosynthesis are amongst such processes. The search for usefulness in the photochemical reactions discovered in rapidly increasing number by the organic chemist often starts - and finishes - with "synthesis". Our first object, therefore, will be to summarize the main areas where photochemistry, in the broadest sense, may be of importance to industry. It will then be apparent that synthesis, _per se,_ is but one such area.

Photo-stabilization and Degradation

The gradual changes in the properties of natural and synthetic materials which occur upon exposure to sunlight are well-known. In the case of fibers, they include modifications in tensile strength, elasticity, and color.

---

1.   G. Ciamician, _Science,_ 36, 385 (1912).

These changes are caused in part by absorption of light and subsequent photochemical reactions.

Chains, on which the fiber strength depends, may be broken, leading to disintegration; new bonds (cross-links) may be formed, leading to brittleness; the reactive species may combine with oxygen with the formation of peroxides which in turn may break down generating more radicals. Color in such fibers, as elsewhere, is usually due to the presence of extended π-electron systems in which low energy electronic transitions may be induced by visible light. These same transitions render the fiber liable to attack by oxygen.

What are the possible solutions to these problems? One approach has been to resolve the difficulty, as do plants, by incorporating into the material substances which act as energy sinks and transform the electronic (light) energy into heat. It will then be a requirement that the incorporation of such a substance shall not modify the desirable properties of the fiber. A start has been made along these lines; one class of such substances used is that of the ortho-substituted benzophenones which have the ability to "waste" electronic energy[2]. These compounds have even been incorporated into **suntan** lotion to protect the "protein polymer" below the applied layer.

In these compounds photoenolization occurs, the enol returns to the ketone in a non-photochemical reaction and a photon has then been converted into relatively innocuous heat[2].

---

2.   G. Porter and M.F. Tchir, <u>Chem. Comm.</u>, 1372 (1970).

An alternative approach is to use an intensely absorbing, but unreactive, material which protects the fiber molecules by denying the light access to them. Obviously such a technique cannot be used generally to stabilize against visible light absorption since this would render the support black.

While photo-degradation is recognized as a multimillion dollar photochemical problem, it has recently been realized that one such type of degradation may be desirable, and may be of assistance in the disposal of the vast amount of solid waste accumulating daily. Very recently[3] efforts have been made to develop materials that will be photolabile enough to disintegrate after a comparatively short exposure to sunlight and such products will soon appear on the market. It is hoped that containers and wrapping materials, particularly those discarded along the highways, can be made of such substances and will then disintegrate to a bio-degradable powder which might even contribute to the soil.

Photochromism

Already in 1911 Ciamician had described the color changes some substances undergo under the influence of light. Included in his historic address[1] were the following words: "Phototropic (now photochromic) substances, which often assume intense colors in the light, and afterwards return in the darkness to their primitive color might be used very effectively . . . The dress of a lady so prepared would change its color according to the intensity of the light . . conforming automatically to the environment: the last word of fashion for the future."

------

3. Chem. Eng. News, July 27, 1970, p. 11; February 8, 1971, p. 52; March 1, 1971; The Christian Science Monitor, August 19, 1970.

Ciamician had great insight - but he could not imagine the large number of possible applications for photochromic susbstances. The breadth of utility becomes greater still because some of these substances - a fact not known to Ciamician - will hold their color permanently until erased by irradiation with light of the color absorbed. Many classes of organic compounds are photochromic, one of particular importance being that of the spiropyrans[4] of which the following is an example.

colorless   colored

Most of the basic research on photochromic substances is being carried out in industry and the information is not readily accessible. Even so, it is clear that great advances have been made in the control of the speed, duration and degree of photochromic action.

One of the most important applications will be data storage and display. For instance, there has been recently

4.   For general reviews on photochromism, see: a) R. Dessauer and J.P. Paris, <u>Adv. in Photochem</u>., 1, 275 (1963); b) W. Luck and H. Sand, <u>Angew. Chem</u>., 76, 463 (1964); c) C. Gliemeroth and K.H. Madler, <u>Ibid</u>., 82, 421 (1970).

announced a computer data display unit using photochromic glass[5]. This photochromic material is a specific formulation of submicroscopic silver halide particles in glass.

Photochromic substances have found application in microimage formation for information storage. For instance, using submicrophotographic technique, 1245 pages of the Bible can be recorded on film of 2 inches area[6]. The greater concentration of data achieved here, in comparison to the process used in ordinary microfilm, is due to the fact that the grain of photochromic material is potentially the size of the molecule itself. The storage of information in this manner has an additional advantage. By scanning with visible light the information at any point can be erased and new information inserted. This is very valuable where a small fraction of stored information needs to be changed often, for example in stock record tabulations.

Amongst other applications are self-darkening glasses for windshields, windows and for sunglasses[7]. The materials now on the market for these applications utilize a photochromic process which has a rather slow response time. But goggles have been developed that darken in microseconds or less when exposed to an intense flash of light.

The applications of photochromism are thus very considerable and would be even greater if the fatigue of the substances involved after millions or more of cycles could be overcome.

Such a wide variety of potential applications, each with unique demands on the photochromic material, will obviously require a number of different systems. Several

---

5.   Chem. Eng. News, April 13, 1970, p. 6.

6.   The National Cash Register Company, Dayton, Ohio.

7.   For example, "Photogray Glass", Optical Products Dept., Corning Glass, Corning, New York; G.P. Smith, J. Photographic Soc., 18, 41 (1970).

classes of organic and inorganic materials exist which have photochromic properties[7], but more are required.

Optical Brighteners

A rapidly developing application of photochemical technology is that of "optical brighteners". This is the conversion of ultraviolet light absorbed (from sunlight or fluorescent lamps) into visible light emitted. Optical brightners now have a practical application in laundry detergents, warning signs, blackboard chalk, clothing, and advertising art. With greater ranges in subtlety of color other uses may be envisaged, and perhaps most dyes or pigments in the future will be emitting in addition to absorbing.

Chemiluminescence

Certain chemical reactions may generate light[8]. Such reactions are found in nature, for instance in the firefly. More recently some of the mechanisms involved have been, at least in part, elucidated.

As an example of a "synthetic" system that for luminol is described below. Many cyclic hydrazides are chemiluminescent when oxidized in the presence of strong base. The light given out is identical with the fluorescence emission of the phthalate dianion. Inexpensive systems with moderate efficiency have been found, so that there is now available an apparatus no more complex than a plastic bag with a division to separate the reactive chemicals[9].

---

8. For reviews on Chemiluminescence see: a) K.-D. Gunderman, _Angew. Chem._, _Internat. Ed._, 4, 566 (1965) b) F. McCapra, _Quart. Rev. (London)_, 20, 485 (1966).

9. For example, "Cyalume" American Cyanamid Company, Organic Chemical Division, Commercial Development Dept., Bound Brook, New Jersey.

The virtues of such light sources, as compared with a flashlight for instance, are, as listed by the manufacturer: lightness, easy storage, a capacity to work under water and lack of sensitivity to shock. A 0.7 fluid ounce package is said to provide detectable light at 2 miles distance with a 3.4 ounce packet being detectable at four miles.

At the present time the intensity of the light is not high and its duration limited to but a few hours. Even so the value of such light sources for emergency in motor vehicles, in mines, for aircraft markers and for convenience in camping, seems clear.

Photopolymerization and Cross-Linking

Several applications have resulted from the combination of photochemical technology to that of polymers (aside from the interest resulting from photodegradation). Applications of photopolymerization and cross-linking include:

solventless coatings for metal or wood[10], solventless, rapid "drying" ink[11], graphic arts, printed coatings, and photomilling. One of the reasons behind efforts to find solventless coatings is the need to restrict the pollution produced by the evaporation of solvents into the atmosphere.

Two fundamentally different approaches have been used. They are those where the irradiation results in bond homolysis of some molecule (polymer, monomer, or initiator) and the subsequent polymerization or cross-linking is free radical in character; or, those where the reaction involves an excited state of the polymer or monomer. There are advantages and disadvantages to either system.

In the first type of system a light-sensitive free radical initiator is frequently added to the monomer. As a result the initiator residue must remain in the polymer, and its presence may adversely affect the properties of the final product. Since this system involves a radical chain mechanism, the reaction may be interfered with by oxygen and so may require a nitrogen atmosphere in the irradiation area.

The second type of system is limited to the cross-linking of polymers where a small amount of cross-linking may affect drastically the properties of the polymer.

One of the first photocross-linking systems used involved the cinnamate containing polymers. The photochemical technology for this innovation dates from the turn of the century[12], and the principle is illustrated below.

---

10.   Technical information is available from Ashdee, 10 So. Tenth Avenue, P.O. Box 325, Evansville, Indiana.

11.   G. Nass, American Ink Maker, January, 1971, p. 25.

12.   D.J. Trecker, Organic Photochem., 2, 63 (1969).

$$-(CH_2 - CH)- \qquad\qquad -(CH_2 - CH)-$$

```
 -(CH₂ - CH)-                          -(CH₂ - CH)-
          |                                     |
          O                                     O
          |                                     |
          C=O                                   C=O
                    hν                  
          CH      ───────►      C₆H₅ ─────┼───── H
          ‖                                     |
C₆H₅CH    CHC₆H₅            H ─────┼───── C₆H₅
   ‖
   CH                              O= C    H
   |                                     |
 O=C                                     O
   |                                     |
   O                               -(CH₂ - CH)-
   |
-(CH₂ - CH)-
```

Applications of such a technique are easily envisaged.
For example, if the non-cross-linked polymer that is not
exposed to the irradiator remains soluble it can be removed.
This leaves the cross-linked polymer covering the irradiated surface; the now exposed surface can then be etched or
the raised polymer can serve as a printing plate. If all
the surface is irradiated then the polymer is 'set' entirely. An application of this aspect is in printing.

Another application of photochemical technology related to photopolymerization and photochromism is in photo-
graphy and office copying. The classic techniques of the
silver halide photographic process and the electrostatic
(Xerox and Electrofax) process of reproduction are being
challenged by organic based systems. Several of the office
copying machines utilize the process of photoconductivity
of inorganic compounds and recently organic photoconducting
systems have been devised.

Synthesis

It is probably in this area of chemical activity that many would predict that photochemistry would have the greatest impact. This may well be the case in the future, but at present, for several reasons, effort and application have been limited.

In comparison with the more familiar thermally activated processes the photochemical technique has certain advantages, and indeed the two processes may often be complementary. The widespread use of photochemical synthesis on an industrial scale implies a need for those products which can conveniently be made this way. There are some few examples which will be discussed later, but first we will consider the economic factors.

It is frequently stated, usually without the benefit of calculations, that only those photochemical reactions which involve long chain free radical processes could be economically feasible; that the cost per Einstein (mole of photons) is too high. The cost of a photochemical process is given by the following equation:

$$\frac{M}{(Kilowatt-hour)} = 0.00302 \times \lambda(A^{o}) \times LE \times \Phi \times FA$$

where:

LE $=$ lamp efficiency

Φ $=$ quantum yield

FA $=$ fraction of emitted light absorbed.

The application of the equation is best understood when illustrated with an example. The quantum yield of benzpinacol obtained by the photochemical reduction of benzophenone in isopropyl alcohol is unity: one photon is required for each molecule of pinacol produced. As an irradiation source we may use a low pressure mercury vapor lamp having, commonly, an efficiency of about 20% for output at

253.7 nm. The apparatus can be designed so that all the light is absorbed by the benzophenone. This would require removal of the product (benzpinacol) which also absorbs weakly at this wavelength. Assuming optimum conditions, then benzpinacol can be produced at 1.49 moles/kwhr, or if the cost of electricity is 0.5 cents/kwhr, the cost of electricity for production of benzpinacol is under 1 cent/lb. In addition the replacement cost of the lamp must be taken into account. High pressure mercury vapor lamps have been developed, having 60 KW capacity where the efficiency falls to 90% of the original after 5000 hours and to about 70% after 10,000 hours[13].

One of the current drawbacks to wider application of photochemical technique is the lack of engineering precedence in this area. Some of the common problems will be enumerated to give a better perspective.

a) Only that portion of the light absorbed by the starting material is useful. If the light source gives a broad band spectrum extending into the visible (high and medium pressure mercury arcs) and the starting material is colorless much energy will be lost.

b) The solvents and reagents must be free of light absorbing impurities.

c) The product and side products must be transparent in the region of irradiation or a procedure for their removal must be devised. This can be a major problem since "tar" frequently coats the surface of the irradiation source.

d) There are presently limitations on light sources. The compact sources have broad band emission and generate much heat, whilst the low pressure sources

---

13.   A.H. Jubb, Ed. in Chem., 8, 23 (1971).

are not compact. Recent developments in plasma-
arc technology[14] suggest that this type of source
may have potential importance, but its use presents new technical problems.

Industrial applications of photochemical synthetic
methods can, for the purposes of this discussion, be divi-
ded into two parts according to the size or type of the
operation.

a) <u>Bulk chemical products</u> require large equipment
and a flow system. Here the low price of the product will
make the process cost a significant fraction of the overall
cost. It is obviously best if the process results in a
large increase in molecular weight from starting material
to product, otherwise all the profit must come from the increased value, pound for pound, of the product relative to
starting material.

b) <u>Fine chemical products</u> may well use smaller
apparatus, similar to the type used for experimental studies. We refer there to chemicals for pharmaceutical,
agricultural, perfumery, flavor and related low-volume
usage. It is here that the most immediate opportunities
are to be found for the application of the considerable
amount of research currently being done in the area of
synthetic organic photochemistry. Indeed, most companies
with this type of business have a growing interest in
photochemistry.

In the first class there are presently three photochemical reactions that are of major importance: photo-

---

14. German Patent 2003132 (July 30, 1970),
Technical information is available from Linde
Division, Union Carbide Corporation, Tarrytown,
New York.

halogenation[15], photonitrosation[16], and photohydrosulfo-
nation[17]. Photohalogenation is well-known and is illustrated, for the chlorination of hydrocarbons. This is a
chain reaction, and consists of the following steps:

Initiation: $\qquad Cl_2 \xrightarrow{h\nu} 2Cl\cdot$

Propagation: $\qquad Cl\cdot + R\text{-}H \longrightarrow R\cdot + HCl$

$\qquad\qquad\qquad R\cdot + Cl_2 \longrightarrow RCl + Cl\cdot$

Termination: $\qquad 2R\cdot \longrightarrow R\text{-}R \;$ or $\; R\text{-}H + R(\text{-}H)$

The photolysis of chlorine is an efficient photochemical reaction and, since chlorine absorbs in the visible
(300 - 500 nm.), high pressure mercury arcs can be utilized
efficiently. Since the reaction has a chain mechanism and
since lengths of over a hundred are not uncommon, as much as
10 - 30 lbs of chlorine per hour can be introduced into a
hydrocarbon when the reaction is initiated with a 400 W arc.
The actual efficiency depends on many factors: phase (liquid
or vapor), concentration, temperature, reaction design, etc.

The second process, photonitrosation, is not as well
understood; in fact, significant progress has been made toward understanding the mechanism of this reaction only in
recent years. It is currently being used to prepare capro-
lactam which in turn is utilized for the production of
Nylon-6[13]. Recently, notice has appeared that Nylon-12

---

15.   a)   Kirk - Othmer, Encyclopedia of Chemical Technology, 2nd Ed., 5, (1964), 85ff;

      b)   M.L. Poutsma, Methods in Free-Radical Chemistry,
1, Ed. E. Huyser, Marcel Dekker, New York, 1969.

16.   a)   M.W. Mosher, N.J. Bunce, Can. J. Chem., 49,
28 (1971);

      b)   E. Muller, Pure and Applied Chem., 16, 153 (1968).

17.   D. Elad, Organic Photochemistry, 2, 168 (1969).

will be prepared by a similar route[18]. Like photohalogenation, photonitrosation involves a free-radical mechanism; however, photonitrosation does not involve a chain
reaction[16]. The quantum yield is of the order of unity;
nonetheless, the process is still economically practical.
An important virtue of nitrosyl chloride which makes the
reaction economically attractive, is the broad light absorption maximum extending throughout the visible spectrum.
The steps involved may simply be represented as shown below:

Initiation: $ClNO \xrightarrow{h\nu} Cl\cdot + NO\cdot$

Hydrogen Abstraction: $Cl\cdot + RH \longrightarrow R\cdot + HCl$

Radical Coupling: $R\cdot + NO\cdot \longrightarrow RNO$

Before the photonitrosation process could become
commercially attractive several engineering problems had to
be solved for which there was no technological precedent:
major achievements were made in lamp design, it was found
that the short wavelength light was responsible for tar
formation and so light absorbing filters were added to the
lamp cooling water; a device was invented which permitted
sulfuric acid scrubbing of the source surface to further
inhibit tar formation. As further experience is gained,
new techniques will evolve to resolve the unique problems
arising in photochemical, as distinct from thermal, processes.

Volatile primary mercaptans find use as additives to
natural gas. They provide the smell which make it possible
to detect otherwise odorless (and dangerous) natural gas.
An industrial process for the preparation of these compounds
is based on the anti-Markovnikov addition of hydrogen sulfide to terminal olefins, a free radical chain reaction initiated by light, the overall scheme being:

18.   Chem. Eng. News, August 31, 1970, p. 12.

275

Initiation: HSH $\xrightarrow{h\nu}$ ·SH + ·H

Propagation: $R-CH=CH_2$ + ·SH $\longrightarrow$ $R-\underset{\cdot}{C}H-CH_2-SH$

$R-\underset{\cdot}{C}H-CH_2-SH$ + HSH $\rightarrow$ $R-CH_2-CH_2-SH$ + ·SH

Termination: Radical coupling and disproportionation

Other free-radical chain reactions are receiving increasing attention[17].

The engineering problems associated with the production of "fine chemicals" are not so severe. Nonetheless
although the number of reactions discovered in the last ten
years is enormous the only reaction (known to us) in use is
an old one: the preparation of vitamin D_2[19]. This is produced commercially by the irradiation of ergosterol contained in fish oils. The relevant steps are shown below:

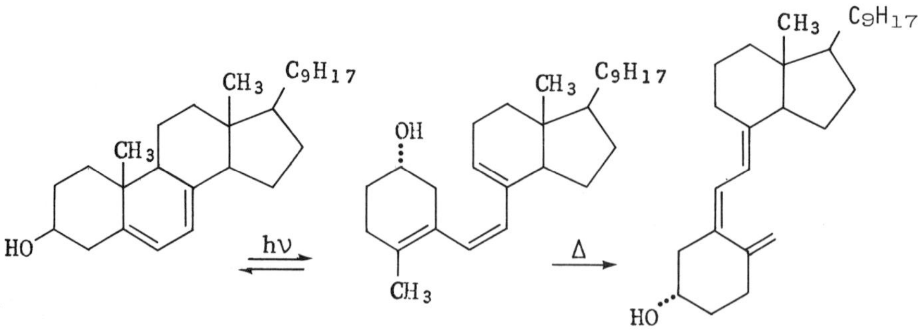

Ergosterol   pre-ergocalciferol   Vitamin D_2

---

19.   S.T. Reid, Progress in Drug Research, 11,
48 (1968).

Photochemical processes are therefore slowly finding application in chemical industry (though not yet to an important degree in synthesis where it might have been expected) and this development should continue. Application to large scale manufacturing must await solution of the technological problems, but the impetus to do this will depend on the growth and interest of chemical industry.

SUBJECT INDEX

279